MARTINUS WILLEM BEIJERINCK

HIS LIFE AND HIS WORK

From the bronze plaque by Professor A. W. M. Odé.

MARTINUS WILLEM BEIJERINCK

HIS LIFE AND HIS WORK

BY

G. VAN ITERSON Jr., L. E. DEN DOOREN DE JONG

AND

A. J. KLUYVER

WITH 13 PLATES OF WHICH TWO COLOURED

Springer-Science+Business Media, B.V.

1940

ISBN 978-94-017-5741-6   ISBN 978-94-017-6107-9 (eBook)
DOI 10.1007/978-94-017-6107-9

Preface

In 1920 a committee was formed by numerous friends and admirers of Martinus Willem Beijerinck with the aim of rendering homage to this great biologist at the occasion of his 70th anniversary. The initiative taken by the first undersigned, who acted as president to this committee, led ultimately to the publication of the "Verzamelde Geschriften" ("Collected Papers") of Beijerinck in five stately volumes.

After Beijerinck's death on January 1st, 1931, it seemed expedient to collect in a final volume those publications of Beijerinck which had appeared after his retirement from the chair at Delft.

On considering the publication of this volume the undersigned arrived at the conclusion that it was most desirable to add to it a detailed biography of the remarkable author of all these memoirs, as well as a comprehensive review of his scientific achievements.

It was then decided that each of the three undersigned should take care of a part of this task. The review was therefore divided into three parts: one, purely biographical, a second part dealing with Beijerinck's studies in the field of general botany, and a third part in which his microbiological work would be considered.

The well-deserved fame which Beijerinck has attained in various branches of biology seems to justify the idea of publishing this biographical and laudatory essay also separately. In doing so it has become possible to make it accessible to a wider circle of readers.

Before finishing this preface the authors wish to express their profound gratitude to all those who have assisted them in their task.

In the first place the precious collaboration of the late Miss H. W. Beijerinck, sister of the scientist, should be most gratefully acknowledged. The liberal way in which she has allowed access to data of biographical interest has been of the greatest value for the successful completion of the purely biographical part. Already during her lifetime, Miss Beijerinck put her diary at the disposal of the second undersigned, a token of confidence which has been highly appreciated. Her unfailing interest in the publication as a whole has greatly stimulated the work. It is a matter of sincere regret to the authors that she did not live to see the book completed. On December 26th, 1937 this energetic and sympathetic woman, whose life was so tightly interwoven with that of her famous brother, quietly passed away at the age of ninety.

The authors also wish to thank Mr. W. M. Beijerinck, retired Major of the Artillery, for information concerning the genealogy of the Beijerinck family.

In the successive phases of the development of the book various British colleagues have been most kind in giving us their advice regarding the linguistic side of the publication. In this respect the authors feel especially, and profoundly, indebted to Dr. Hugh Nicol, bacteriologist of Rothamsted Experimental Station, for the untiring and devoted way in which he has accomplished the most unselfish task of correcting the manuscripts from the point of view of the language. In doing so, he has not only eliminated numerous shortcomings in English style and composition, but at several places his critical suggestions — which were always to the point — have greatly influenced the redaction of the survey given.

Delft, October 1940.

G. van Iterson Jr.
L. E. den Dooren de Jong
A. J. Kluyver

Contents

PART I. BEIJERINCK, THE MAN

by

L. E. den Dooren de Jong

PART II. BEIJERINCK, THE BOTANIST

by

G. van Iterson Jr.

PART III. BEIJERINCK, THE MICROBIOLOGIST

by

A. J. Kluyver

APPENDICES

List of Plates

PART I

BEIJERINCK

THE MAN

(March 16th, 1851 – January 1st, 1931)

BY

L. E. DEN DOOREN DE JONG

CHAPTER I

DESCENT

According to recent genealogical researches [1]), the BEIJERINCK family seems to come from Twente, a region in the Netherlands province of Overijsel, where at any rate since 1429 at Tilligte near Oldenzaal two farms are situated, "Beyerinck" and "Olde Beyerinck", which most likely were their "incke" [2]). Of old, BEYERINCKS have lived on these farms [3]). Among others a certain Johan is mentioned in 1558 as inheriting a farm near Hengelo.

These BEYERINCKS were and still are Roman Catholics. Probably some went to Amsterdam and afterwards to Kampen, but this branch has died out. Another branch went to the Achterhoek, the eastern part of the province of Gelderland, and of that branch, which became Protestant, the genealogy is completely known [4]).

JORDEN BEYERINGS, on April 13th, 1628 attended the Lord's Supper at Doetinchem with his wife Aeltjen. One of his sons, PETER BEYERINCK, was married to DEUKEN FRANSSEN, their son Jorden was baptized at Doetinchem on May 29th, 1659. This Jorden (or Jordan) was a weaver, went to Nijmegen and was entered there on May 3rd, 1682 as a citizen, as is testified in the following resolution of the council: "JORDAN BEYERINCK, geboortigh van Doetinchem, van de waare Christelijke gereformeerde Religie synde, is tot borger deser stadt aangenomen, mits betalende het recht daartoe staande *et praestitit juramentum*" [5]).

Destitute descendants of this Jordan have the right, when 60 years old, of applying to the old City Almshouses of Nijmegen for admission or other relief.

On May 21st, 1683 Jordan was married to ANNA CATHARINA VAN JUCHEN. One of their sons, Peter, was born at Nijmegen on May 16th,

---

[1]) See VAN DOORNINCK's register in the old provincial record office of Overijsel; Vol. III, p. 12 (1424–1456) and Vol. IV, p. 128 (1456–1496).

[2]) An "incke" (in modern Dutch: enk) is the name of a part of arable land, as a rule situated somewhat higher than its surroundings. As appears from the names, the origin of many villages in Holland may be traced back to these "inckes" or "enckes".

[3]) In one of the houses there is still a beam on which is written: "1653, 5 April. JAN TER LINDE ende JENNE BEYERINCK".

[4]) See for this: Nederland's Patriciaat Anno 1919, 10th Vol. pp. 9–21 and the genealogical register of the BEIJERINCK family (S. J. VAN AMERONGEN, Amersfoort).

[5]) Translated: "JORDAN BEYERINCK, born at Doetinchem, of the true Christian reformed Religion, has been admitted as a citizen of this town, provided he pays the tax raised for this *et praestitit juramentum*".

1684, and married on February 11th, 1714 LEENDERTJE CRANE. He was a surveyor by profession. One of their children was Martinus, born March 31st, 1718 at Nijmegen (Cf. Plate II). He too was a surveyor and moreover a municipal official of Nijmegen. On April 12th, 1750 he was married to GIJSBERTA SWINNAS. From this marriage 3 sons, Willem, Leonardus and Frederik, were born.

Frederik was the grandfather of Professor BEIJERINCK. He was born at Nijmegen in 1766, and afterwards occupied the important post of chief engineer of the Department of Buildings and Roads, being entrusted with the survey of the rivers Rhine and Waal, as far as they ran in Gelderland. For this purpose he lived alternately at Nijmegen and Arnhem. The government acknowledged his merits by knighting him in the order of "de Nederlandsche Leeuw". A portrait of Frederik showing an undeniable resemblance to Professor BEIJERINCK is in the possession of the family, and is reproduced in Plate II. Frederik was married twice. Firstly to ELISABETH REIJNEN, from which marriage issued:

1) Martinus 1803–1879.
2) Derk, father of Prof. BEIJERINCK 1805–1879.

After the death of his first wife he married JACOMINA CRIJNEN, who gave birth to two more sons:

1) Leonard Willem [1]).
2) Willem Cornelis.

FREDERIK BEIJERINCK died in 1838.

Although the straight line of descent is left here, it may be mentioned that Martinus, uncle of Professor BEIJERINCK, born in 1803, had a much more successful career than BEIJERINCK's father whose misfortunes are related in the following pages. MARTINUS BEIJERINCK started his career as an engineer of the Department of Buildings and Roads, and afterwards became a professor at the Polytechnical School of Delft.

We now arrive at Professor BEIJERINCK's parents. DERK BEIJERINCK, his father was born April 21st, 1805, baptized May 19th, 1805 at Nijmegen, and died on January 22nd, 1879 at Elst, Over-Betuwe. Derk (Cf. Plate II) was married on April 27th, 1843 to JEANNETTE HENRIËTTE VAN SLOGTEREN (born at Hoorn November 29th, 1811, died April 16th, 1875 at Elst) (Cf. Plate II). She was the daughter of the Rev. JOHANNES VAN SLOGTEREN, linguist and minister first at Keppel and Doetinchem, and afterwards at Hoorn. From this marriage were born:

1) Frederik Leonard, born at Amsterdam, November 26th, 1844, died at Almelo December 29th, 1883.

[1]) This Leonard Willem had a rather remarkable career. As an equerry and at the same time a great friend to Bernhard, Duke of Saxen-Weimar (Commander in Chief of the Dutch Indian Army) he visited the Indies twice. On the second journey (in 1849) along the Isthmus of Suez the Duke and he were the guests of the Viceroy of Egypt, Abbas Pasha.

Martinus Beijerinck (1718–1782);
great-grandfather of the scientist.

Frederik Beijerinck (1766–1838);
grandfather of the scientist.

Derk Beijerinck (1805–1879);
father of the scientist.

Jeannette Henriëtte van Slogteren
(1811–1875); mother of the scientist.

2) Henriëtte Wilhelmina, born at Amsterdam, February 23rd, 1847, died at Gorssel December 26th, 1937.
3) Johanna Hermana Alida, born at Amsterdam, February 2nd, 1849, died at Gorssel September 24th, 1923.
4) MARTINUS WILLEM, born at Amsterdam, March 16th, 1851, died at Gorssel January 1st, 1931.

When examining the collateral branches of the BEIJERINCK family, we find that a striking proportion of its male members have occupied official and sometimes important posts. There is a large number of civil engineers, inspectors and chief inspectors of the Department of Buildings and Roads, officers and field officers in the East Indian army (among whom one knight M.W.O. [1])), East Indian civil servants, *e.g.*, residents, civil officers for taxes and registration, surveyors, etc. From all this it is evident that intelligence and trustworthiness are inherent in the BEIJERINCK family. However, nothing in this pedigree seems to indicate the appearance of a character like that which imbues the subject of this biography.

---

[1]) Militaire Willems Orde, the Netherlands' equivalent of the Victoria Cross.

CHAPTER II

CHILDHOOD

Before proceeding to a description of MARTINUS' childhood, we should say something about DERK BEIJERINCK and his family. From private communications we have learnt that Derk had a cheerful, strong and brave nature. Like his father, Derk had artistic gifts which are also apparent in his children. MARTINUS' sister Henriëtte, for instance, made several drawings and pictures of plants and microbes, which are still used for teaching purposes in the laboratories for Microbiology and for Botany at Delft.

When five months old, DERK BEIJERINCK had the misfortune to lose his mother, ELISABETH REIJNEN (probably a daughter of a chemist at Nijmegen).

If Derk's mother had remained alive, her youngest son would no doubt have got on better than he did. The second wife of FREDERIK BEIJERINCK did not take to the children of his first marriage, and neglected them. Martinus managed to overcome the difficulties engendered by the home atmosphere, and became, as has already been mentioned, professor at the Polytechnical School at Delft. When still too young to decide his own vocation the intelligent and quick-witted Derk was forced to go into business for which he was given no training, and had no talents. In 1830 he volunteered, and went through the campaign against Belgium. After this he was given the option of going to the Indies with the rank of second lieutenant, or of remaining in the army in this country in a lower rank. Derk, however, preferred to retire from the army, and received the volunteer's cross.

From his mother's inheritance his father then bought for him a tobacco business at Amsterdam, *viz.*, 81 Op het Water (Damrak), called "Het Wapen van Oldenburg" ("The Oldenburg Arms"). On April 27th, 1843 he married JEANNETTE HENRIËTTE VAN SLOGTEREN.

Being conducted in defiance of sound principles, the business slowly collapsed, in spite of all Derk's well-meaning efforts, and had to be sold in 1853. The sale left him with only a small sum, since the money brought by the mother had been sunk in buying the shop.

Consequently it was into a family suffering from financial difficulties that on the Sunday morning of March 16th, 1851 MARTINUS WILLEM BEIJERINCK was born as the last of Derk's four children. The others were Frederik, Henriëtte, and Johanna, then 6, 4, and 2 years old respectively. When MARTINUS was two years old, the family moved to Naarden, where life was less expensive than in Amsterdam. Here strenuous efforts were made to find a position. On the recommendation of a friend Derk obtained a situation as clerk in the Haarlem booking-office of the Hollandsche IJzeren Spoorweg Maatschappij [1]), so that at the beginning of 1854 the family had to break up again.

They found a suitable house in Haarlem at the end of St. Janstraat, not far from the Lage Bolwerk. It had a fairly large garden in front and at back. Here Derk in his spare time grew vegetables, for he was fond of gardening, and was a great lover of nature. During the many walks he took with wife and children on his few free days the eyes of his youngest son were no doubt opened to the beauty of nature, which afterwards became his alpha and omega. Also in Haarlem Derk had a hard life. The exigencies of the railway service resulted in Derk's hours being long and irregular, and frequently his working day extended from half past six in the morning until half past ten at night with variable periods off duty in between. With such opportunities as his scanty leisure afforded he still found time to teach the subjects of the elementary school to the children for whom he could not afford schooling. To the three R's he added French, English, a little German, drawing, and the elements of cosmography and physics. In this way the BEIJERINCK children were educated, and, when afterwards they went to school, they were not behindhand. The dear, gentle, yet equally energetic mother taught her daughters needlework and housekeeping.

MARTINUS as a boy was sensitive, and kind, with a strong sense of justice. If during play with his sisters he happened to fall and hurt himself, and the mother thought that he had not been looked after properly, he always said "They could not help it", for fear that they would get into trouble.

In spite of the greatest economy, the house was too big, and at last the family went to live in a workman's cottage which, although newly built, was poky and inconvenient. Wife and husband had seen better days, and were thoroughly miserable in the new house. More trouble came when, some time later, the husband fell ill, but both bore up bravely.

In 1859, when MARTINUS was eight years old, his father was transferred to Leiden, where he got a post in the goodsoffice of the same railway, and where he could make use of his knowledge of English, French and German. However, his situation there was far from pleasant. His immediate superior was a former coachman who browbeat the better-bred man, and lost no opportunity of asserting himself at his expense.

For the children the four years at Leiden were very pleasant. They now had a better house, situated on the Mare [2]) at some distance from

[1]) Holland Railway Company.
[2]) A water course.

Leiden, but the family was occasionally attacked by malaria. Derk, who was still ailing, was nevertheless able to devote himself once more to the education of his children, in which occupation he was sometimes assisted by the eldest son, Frederik. As befitted a minister's daughter, the mother gave her children a Christian upbringing. On Sunday mornings the father used to read to them from a translation of HEINRICH ZSCHOKKE's "Stunden der Andacht" ("Devotional Hours") which made a lasting impression upon them. Every morning, too, the mother used to read something to them from the Bible, and made the children learn psalms and hymns by heart. She herself went to church frequently.

It is worth mentioning that when MARTINUS was 10 or 11 years old, he was subject to fits; for some time he was so seriously ill that his parents feared for his life. In those years a small legacy from an aunt of MARTINUS' mother brought a considerable relief. The family was a little better off now. They had never lived above their income. The mother, the soul of the family, had by her good housewifely management avoided getting into debt, but at the same time she had seen to it that the children should not go short of necessities. They had never actually suffered want, and indeed they had no real notion of the cares that weighed down their parents. But the financial difficulties of the parents prevented them, although people of culture and good standing, from having that intellectual contact with the outer world which would have assisted the social development of their children. In all probability this contributed to the inclination to solitariness of the youngest son who, like his sister Johanna, was fundamentally gentle and timid. Frederik, the eldest of the four, was a sturdy boy; and on the lonely winter evenings, when the father was at his office, the mother was always glad if Frederik was at home. Frederik was an intelligent lad, but, through lack of means, did not get the best training. Yet, when he was 18 years old, and had to join the army as a conscript, the family managed to take a substitute for him. Quite early he was apprenticed in the office of a surveyor in order to qualify for admission to the cadastral survey, and later he came to be a surveyor.

In 1863 Derk was transferred back to Haarlem, and the family went to live at the Nieuwe Gracht overlooking the Spaarne, Koude Hoorn and Scheepmakersdijk. The children then were 18, 16, 14 and 12 years old respectively. Frederik was training for the assistant surveyor examination, Henriëtte became a pupil-teacher, Johanna went to school and studied for the elementary school teacher's examination, and MARTINUS attended the elementary school of Master KNOOP and subsequently the "Hoogere Burgerschool" (secondary school) at Haarlem, where Dr. E. VAN DER VEN was head-master. Few recollections of that period remained with Professor BEIJERINCK in later life. All he remembered was that it had been a miserable time for him. In the elementary school the master once called upon him to tell the class something about JACOBA OF BAVARIA. As he appeared to know nothing about her, a boy whispered to him: "Say something about the jugs!" [1]). It is amusing that on telling this story in later years the then professor added to this: "I had already learnt to detest history". In the second form of the secondary school he once wore a green jacket which, probably owing to the family poverty, was rather old-fashioned. The boys laughed at it, and MARTINUS took it to heart. It is not unlikely that such continual teasings contributed to the fact that MARTINUS in later years was mostly gloomy and reserved.

During his school days MARTINUS associated with older people. Amongst these, special mention should be made of Mr. FREDERIK WILLEM VAN EEDEN [2]), a well-known botanist who did a great deal to arouse interest in the flora of the Netherlands at home and overseas, and who ultimately rose to be Director of the Colonial Museum at Haarlem.

MARTINUS had the great privilege of taking many botanical walks in the surroundings of Haarlem with VAN EEDEN, and it seems extremely probable that it was this naturalist who aroused his interest in plants and animals. He also made several excursions with Mr. KNIPSCHEER, an older gentleman who formerly held the high position of resident in the Netherlands Indian Civil Service, and whose grandson Hendrik went to the same school as MARTINUS did.

It was also at this school that young BEIJERINCK got to know two boys, LEO and CAREL DE LEEUW, whose parents, Mr. and Mrs. DE LEEUW–PENNINCK HOOFT, lived in the Anna Paulowna polder. The reclaiming of this polder had been carried out owing to the initiative taken by Mr. DE LEEUW who fittingly became its first dike-reeve and major. MARTINUS often enjoyed the hospitality of the family. Here he also made friends with the daughter, AMY DE LEEUW, who later became well-known under her pen-name GEERTRUIDA CARELSEN.

This friendship continued for long years, and was based on their common love for flowers and plants. During BEIJERINCK's visits to Anna Paulowna they used to study the development of the flora of the new land, and many times made botanical excursions to the island of Wieringen. In these years also HUGO DE VRIES was at several times a guest at the Anna Paulowna-polder house.

In this period of his life BEIJERINCK is described as having a gift of application coupled with a steady nature, and since he was also pleasant and witty, his people were very fond of him. It is noteworthy that a cousin at the beginning of his secondary school

---

[1]) JACOBA OF BAVARIA, Countess of Holland (1401–1435), is a notorious figure in the history of the Netherlands. She lived for some time at the castle of Teilingen near Haarlem. Afterwards many jugs have been found in the castle moat. They are supposed to have been thrown therein on the occasion of the festivals organized by Jacoba.

[2]) Father of FREDERIK VAN EEDEN, famous Netherlands man of letters and sociologist.

career was dubious about his intellectual capacities, thinking he was a dunce because he had some difficulty in learning the tenses of French verbs. It later became apparent that he could profit considerable from study, and he usually was second or third in class.

At first his health was very indifferent. When 13 years old, attacks of fever confined him frequently to bed. When he was 14 he showed signs of heart weakness. After this age his health improved a great deal.

In 1866, in connection with a competition instituted by the "Hollandsche Maatschappij van Landbouw" [1]), of which the well-known J. H. KRELAGE, then was president, he began to make a herbarium of 150 kinds of plants found in the surroundings of Haarlem. Only young people under the age of 16 were allowed to compete; each plant had to be given its Dutch and Latin names, and the date and place of finding had to be stated. The first attempt was a failure; the plants were not well dried, and some became mouldy, so that he had to start again. The second effort failed likewise, and MARTINUS was totally disheartened. But, after his mother had encouraged him, he began anew, and this time he mastered the technique. He began to take pleasure in it, and said: "Whether I get a prize or not does not matter, but I'll stick to botany". Nobody guessed then how much truth this statement contained.

The collection was sent in, and the 15-years-old MARTINUS obtained the first prize, consisting of the silver medal of the "Hollandsche Maatschappij van Landbouw", with his name engraved in it, and also the valuable "Flora van Nederland" by C. A. J. A. OUDEMANS, with atlas.

---

[1]) Netherlands Agricultural Society.

CHAPTER III

ADOLESCENCE

Martinus worked on quietly, and in 1869 he passed the final examination of the Hoogere Burgerschool (secondary school). Although he was very much afraid of this examination, he did very well. The distribution of the certificates took place in the "St. Janskerk" at Haarlem.

Meanwhile the state of affairs in the Beijerinck family had become more difficult again. Derk Beijerinck had to retire because he had reached the age limit. His pension was a very modest one, so the family had to reduce expenses even more. Fortunately, relief came in two ways. Frederik had become a cadastral surveyor, and suggested that the family should come and live with him in den Briel, while Martinus by the generous support of an uncle on the mother's side, A. L. van Slogteren, notary at Enkhuizen, was enabled to study technology at the Delft Polytechnical School. At the time this course of study took only three years, whereas University training took twice as long. Although the decision can be understood from a financial point of view, yet it seemed at first sight regrettable, considering Martinus' pronounced botanical inclinations, that he was not allowed to take up his favourite subject straight away. Nevertheless, his later career shows that these years of study at Delft yielded fruit. A mere botanist would never have had the deep chemical insight into microbiological processes which the later professor had. A great part of the publications from the Laboratory at Delft are, indeed, concerned with subjects on the border-line between biology and chemistry.

According to a personal communication by Professor Beijerinck the practical training of technologists at the Polytechnical School in the years 1869–1872, when he studied at Delft, was extremely poor. It was very rarely that the professor of chemistry came into the laboratory! It was usual among the undergraduates to work there for about a week every six months. However, what is important is that Beijerinck at that time formed a great friendship with J. H. van 't Hoff, the later Nobel-Prize laureate in chemistry, and who was then also studying technology. They lived together at the Camaretten, and had great trouble about their food, which was bad and expensive, so that finally they put themselves on a ration of rice and beefsteak. In order to satisfy their longing for experimentation Beijerinck and van 't Hoff made many chemical experiments in their rooms. Once they boiled dead moles with caustic soda, freed the skeletons, and then treated them with hydrochloric acid with the aim of preparing glue from the bones. This resulted in the landlady giving them notice to quit.

It has to be admitted that a good deal could be learned from the theoretical teaching at the Polytechnical School, and BEIJERINCK did work hard at this part of his studies. Sundays being devoid of lectures were very lonely days for him. It was only occasionally that he could afford to go and see his people at den Briel, sometimes together with VAN 'T HOFF, who generally spent his Sundays at Rotterdam. BEIJERINCK was a melancholy lad in those years, and when the final examination approached he became even more depressed than usual.

However, on July 5th, 1872 he passed brilliantly, but started worrying at once how to obtain a post. For this purpose he answered an advertisement of the Minister of the Colonies who was appointing three young men with the secondary school certificate to study in Prussia at the expense of the State for the Forest Service in the Netherlands East Indies. BEIJERINCK had an interview with Minister FRANSEN VAN DE PUTTE, and obtained his promise of the vacant post, provided he satisfied the medical examiners. To his great distress, however, he was not accepted because of an assumed heart weakness. "He might stay alive here, but in the Indies he would develop heart trouble within two years", was the opinion of the medical authorities.

We do not know which were the circumstances that enabled BEIJERINCK at last — after losing three years, as he expressed himself later — to follow his inclination, and to start the study of biology at the University of Leiden. On October 23rd, 1872, he placed his name on the books of the University and set to work with great diligence. Being already well trained theoretically, he was able to pass the candidate examination after eight months. Minister THORBECKE had given him, as well as VAN 'T HOFF and HUBRECHT [1]), special exemption from matriculation, so that he could study at Leiden without having the classical education that was then requisite. The certificate of this dispensation got lost on the day before the examination, and in despair BEIJERINCK went to the Minister of Home Affairs, DE GEER, who sent him another copy that same night. Afterwards one of his friends helped him to look for it, and found the original document behind the mirror of a dressing-case. On the day of the examination, therefore, he possessed two copies. On June 7th, 1873 he passed the candidate examination *magna cum laude*. He immediately applied for the post of teacher at the secondary school in Wageningen, unsuccessfully however.

---

[1]) The later well-known professor of embryology at the University of Utrecht.

Frederik Leonard Beijerinck (1844–1883); brother of the scientist.

Henriëtte Wilhelmina Beijerinck (1847–1937); elder sister of the scientist.

Johanna Hermana Alida Beijerinck (1849–1923); younger sister of the scientist.

M. W. Beijerinck as a student, at the age of 20.

CHAPTER IV

THE SECONDARY SCHOOL TEACHER

Only two months later a telegram arrived from the burgomaster of Warffum — a small town in the province of Groningen — notifying BEIJERINCK of his appointment as a teacher at the Agricultural School there on a salary of *f* 1800.— a year. BEIJERINCK did not at all like the idea of going to Warffum, but things turned out better than he had expected. The class rooms were good. For the training of 20 young men from the peasantry there were 9 teachers with whom, however, BEIJERINCK did not always get on very well. Groningen being fairly near, the post has the advantage of giving an opportunity for University studies. Here BEIJERINCK had his name entered in the very same year. In September 1874, however, the Agricultural School was done away with for a year and to his indignation BEIJERINCK was dismissed on January 1st, 1875 with four weeks' notice. However, he received a part-time post as a teacher at the State secondary school at Warffum on a salary of *f* 200.—, and had moreover *f* 1000.— as half-pay.

Meanwhile his parents and sisters had gone to live at Elst in the province of Gelderland, where his father in 1872 had been made caretaker of the "Ingelandshuis" [1]). Portraits of BEIJERINCK, his brother, and his sisters at this period are reproduced in Plate III.

In 1875 the family met with the misfortune of losing the mother. MARTINUS and his brother Frederik arrived just too late to see her still alive.

About that time BEIJERINCK had much trouble with his health. A consultation with a medical professor did not bring any organic defects to light; all the troubles were put down to nervousness [2]).

By the good offices of his fatherly friend F. W. VAN EEDEN, BEIJERINCK next had the chance of being appointed as steward of the country-seat Elswoud near Haarlem on a salary of *f* 1200.— a year, but he decided to keep to scientific work, and began to prepare for the "doctoral examination", for which purpose he visited Professor SURINGAR at Leiden. In June 1875 he wrote to his father and sisters that he had been admitted to the third part of the "doctoral examination", which meant that now he might take his Doctor's degree. Typical of

1) "Landholders house".
2) In later life also BEIJERINCK was always worried about his health, and often tormented himself with thoughts of imaginary diseases.

his low state of mind is the expression in his letter, "as little as I was worried about it beforehand, as little am I happy now that it is over".

A few months afterwards BEIJERINCK became a teacher at the "Hoogere Burgerschool" in Utrecht. He would have preferred to teach at the "Hoogere Burgerschool" in Amsterdam, whence HUGO DE VRIES had just resigned, but a kind of diffidence kept him from applying. In the old cathedral town of Utrecht, he took rooms above the Swiss shop, and had the great advantage of again coming into touch with his friend VAN 'T HOFF, who was then an assistant in chemistry at the Veterinary School [1]). They often had dinner together. BEIJERINCK had about 100 pupils who not seldom gave him trouble; some of them he had to send out of the class-room as a disciplinary measure. Though in later years, BEIJERINCK liked teaching, he took little pleasure in it at this period. He endured a lonely life, since, owing to his self-sufficient nature, he did not seek company. It is remarkable that his brother Frederik, who likewise was very intelligent, was totally different in this respect. Yet, BEIJERINCK was not quite so forlorn as it might seem. His sisters and father were very fond of him, were proud of their clever brother and son, and helped him as much as they could. They sent him extra provisions, and even drinking-water, from Elst (!), because he imagined that the "bad" drinking-water at Utrecht had affected his health. They also sent him on his request various plant-galls. In October 1875 he decided to take these teratological formations as the subject of his doctorate thesis.

In the summer of 1876 BEIJERINCK developed an inflammation of a rib and became seriously ill [2]). His friend VAN 'T HOFF nursed him carefully and regularly wrote to Elst, from where his father came to see him from time to time. During his illness he received the news of his appointment as teacher at the Agricultural High School in Wageningen at a salary of *f* 1800.— a year. BEIJERINCK was very pleased with it, for now he could exclusively teach his favourite subject, botany. In the autumn, when he had recovered, he entered upon his duties at Wageningen.

At that time JONGKINDT KONINCK was Director of the Agricultural High School; the pupils were farmers' sons and rich young men, many of whom had not distinguished themselves at other schools. The majority were boarders.

In the beginning of 1877 BEIJERINCK's first important paper, written in Utrecht and entitled "Ueber Pflanzengallen", was published in the "Botanische Zeitung". It was rather severely criticized by SNELLEN VAN VOLLENHOVEN, and the criticism greatly disheartened BEIJE-

[1]) According to a personal communication by Professor BEIJERINCK, VAN 'T HOFF was highly indignant with the Emperor of Brazil who — when visiting this school — took him for an amanuensis.

[2]) It does not seem unlikely that this was an unrecognized case of typhoid fever which BEIJERINCK may well have contracted from the water of the rural supply of Elst!

BIJDRAGE

TOT DE

MORPHOLOGIE DER PLANTEGALLEN.

ACADEMISCH PROEFSCHRIFT

TER VERKRIJGING VAN DEN GRAAD

VAN

DOCTOR IN DE WIS- EN NATUURKUNDE,

AAN DE HOOGESCHOOL TE LEIDEN,

OP GEZAG VAN DEN RECTOR MAGNIFICUS

Dr. P. VAN GEER,

HOOGLEERAAR IN DE FACULTEIT DER WIS- EN NATUURKUNDE,

op DONDERDAG den 14den JUNI 1877, des namiddags te 3 uren,

IN HET OPENBAAR TE VERDEDIGEN

DOOR

MARTINUS WILLEM BEIJERINCK,

GEBOREN TE AMSTERDAM.

UTRECHT
Firma L. E. BOSCH EN ZOON.
1877.

[Facsimile of title page of Beijerinck's thesis for the degree of Doctor of Science]

RINCK. Professor SURINGAR, however, put fresh courage into him, and allowed him to take his Doctor's degree on the work in the paper.

On Thursday June 14th, 1877 the promotion ceremony took place. BEIJERINCK would have preferred to do it privately but, as his name had not been on the books of Leiden University for the last two years, it had to be done in public. His dissertation was entitled: "Bijdrage tot de Morphologie der Plantegallen", and was dedicated to his father (Cf. Plate IV). It was accompanied by 20 "stellingen" which have been reproduced in Appendix A, since they are representative of the scientific outlook of BEIJERINCK in the first stage of his development. On reading these "stellingen" one is struck by the briefness of many of them (Cf. I, III, V, VII, VIII, XI, XV, XVI, XVII, XIX and XX), and also by the resoluteness in which they were drawn up. To say things briefly and concisely was a quality which marked BEIJERINCK throughout his career. "A discovery is great when one can communicate it in passing" was one of his favourite sayings. It was not his way to take an intermediate standpoint in scientific matters; BEIJERINCK liked pithy statements, and nevertheless he was not seldom right!

A second point which draws the attention in these "stellingen" is BEIJERINCK's versatility. Besides subjects from the most divergent domains of biology, physical and chemical items apparently attracted BEIJERINCK's interest (Cf. I, II, III, IV and XX). One of the "stellingen" (IV) testifies to his close relation to VAN 'T HOFF. Some of them refer to the cosmos (I, XX). This many-sidedness has characterized BEIJERINCK till the moment of his death. "Stelling" X is devoted to DARWIN about whom BEIJERINCK always spoke with the greatest admiration.

Although BEIJERINCK was very worried about the promotion ceremony, all went off quite well. As was usual at the time, the *promovendus* in black with white gloves drove in a carriage and pair to the great hall of the University, and there joined the procession of beadle, professors, and opponents. The latter were VAN 'T HOFF and VAN RENESSE. At 5 o'clock the Latin speech of Professor SURINGAR was finished, and BEIJERINCK obtained the first degree. Since he had never learnt Latin, he did not understand a word of the speech, and he bowed before the end. The customary graduation dinner was not given, for BEIJERINCK could not afford it.

As a teacher at the Agricultural High School BEIJERINCK was in his element. This period, or at least the beginning of it, was in many respects the happiest of BEIJERINCK's life. That he could entirely devote himself to botany is proved by the great number of articles, often of considerable length, which he wrote at that time. All articles of the first volume and the first four of the second volume of the Collected Papers were written there. The article "Onderzoekingen over de besmettelijkheid der gomziekte bij planten" [1]) was communicated to

[1]) "Researches on the contagiousness of the gum disease in plants".

the Amsterdam Academy in 1883 by Professors DE VRIES and RAUWENHOFF. As demonstration material BEIJERINCK added to the manuscript a branch showing gum formation as the result of one of his infection experiments.

In May 1884 BEIJERINCK was elected a member of the "Koninklijke Akademie van Wetenschappen" at Amsterdam (Royal Academy of Sciences); soon afterwards he was installed. Once in the Academy he came into regular contact with prominent scholars of that time, such as HUGO DE VRIES, C. A. J. A. OUDEMANS, F. C. DONDERS and TH. W. ENGELMANN. In later years he came into close touch with the physiologist C. A. PEKELHARING. At Wageningen, together with ADOLF MAYER, he founded the "Natuurwetenschappelijk Gezelschap", a society for the encouragement of the natural sciences.

On January 22nd, 1879 his father died, and he then went to live with his sisters Henriëtte and Johanna in the Dijkstraat at Wageningen. This should have made his hitherto solitary life more agreeable. Yet BEIJERINCK remained more or less gloomy, as appears from the following complaint found in the diary of his sister Henriëtte: "On our walks he often remains silent for hours, which always makes me slightly annoyed with him". Henriëtte, then holder of the teaching-certificate for drawing in elementary schools, helped her brother a great deal in drawing beautiful botanical pictures for teaching purposes. Sister Johanna, a school-teacher, assisted him to translate his articles, particularly into English. The trio led a rather lonely life, and mixed with very few people in Wageningen. However, BEIJERINCK came on friendly terms with his colleague OTTO PITSCH and with Dr. M. KREUNEN, a teacher of the classical languages at the Gymnasium. On Sundays they often made long walks together [1]).

Professor MOLL from Groningen and particularly VAN 'T HOFF — not yet 30 years old and already a professor in the University of Amsterdam — sometimes came to see BEIJERINCK and his sisters.

In later years also the family received much friendship from VAN 'T HOFF and his wife, Mrs. J. VAN 'T HOFF–MEES. It was owing to them that Henriëtte was enabled to continue her studies at Amsterdam in order to work for the teaching-certificate for drawing in secondary schools. With Henriëtte away during the latter years of his residence at Wageningen, BEIJERINCK was left with only his sister Johanna. A deep sorrow was caused by the tragic death of their brother Frederik on December 29th, 1883. BEIJERINCK was greatly distressed by this bereavement.

In other repects also his life at Wageningen was getting less pleasant. His standing with the director of the Agricultural School

1) The friendship with Dr. KREUNEN lasted till the latter's death. There is no doubt that Dr. KREUNEN rendered BEIJERINCK numerous valuable services. When new microbe species had to be named BEIJERINCK sent a brief description of the more typical properties of the organism to KREUNEN who then proposed an appropriate Latin name.

failed to improve. It is hard to tell whose fault this was, but it is quite certain that a quick-tempered, arbitrary man like BEIJERINCK, who never minced his words, must with his conscious intellectual superiority have been very difficult to get on with. He also found the students very trying, they often gave cause for complaints; one of them was sent down at his request.

Owing to the high standard of his publications, lectures, and scientific reports BEIJERINCK was already held in high esteem. This led a far-seeing industrialist J. C. VAN MARKEN, the director of the "Nederlandsche Gist- en Spiritusfabriek" [1]) at Delft, to invite BEIJERINCK in the autumn on 1884 to accept the position of bacteriologist at this factory. A salary of *f* 4500.—, which was quite high for that time, was offered to BEIJERINCK, and besides he was also promised a new laboratory. VAN MARKEN left him until January 1st, 1885 to decide. This attractive offer greatly embarrassed BEIJERINCK, who always had difficulty in taking important decisions. His friends, HUGO DE VRIES and VAN 'T HOFF, were called upon to give their advice. When, in the beginning of December 1884 BEIJERINCK decided to accept the new post, his admirers at Wageningen, particularly the staff of the Agricultural School, made it even more difficult to him by sending a petition to the government praying for him to be retained at the School. Professor SALVERDA suggested a salary of *f* 3500.— and a new laboratory in the garden of the school. Time was getting on, the government made no move, and on December 31st BEIJERINCK accepted the post at Delft.

As the laboratory of the factory was not then finished, BEIJERINCK went abroad in order to prepare for his new task. On the programme were visits to the laboratories of DE BARY, KOCH and HANSEN.

His first visit was to DE BARY at Strasburg, at whose laboratory a more or less awkward incident took place [2]). BEIJERINCK, whose scientific enthusiasm and fondness for dispute knew no bounds, appears to have pointed out errors with so much vehemence that DE BARY asked him to keep his knowledge to himself.

In later years BEIJERINCK used to say that in HANSEN's Laboratory at Copenhagen he was fobbed off with trifles, a statement which can hardly be considered to give a fair idea of his experiences. Since he expected to learn even less from KOCH (!) he had given up the projected visit to Berlin [3]).

In April 1885 BEIJERINCK paid a last visit to the Agricultural School, receiving many marks of appreciation. In June he made a journey to Basle with HUGO DE VRIES, and in September 1885 he entered upon his functions at the yeast factory at Delft.

---

[1]) "Netherlands Yeast and Spirit Works".
[2]) This incident was later reported to the author by Professor BEIJERINCK himself.
[3]) During his whole lifetime BEIJERINCK showed a rather pronounced dislike for medical bacteriology.

CHAPTER V

THE INDUSTRIAL MICROBIOLOGIST

This period — however important it may have been from a scientific point of view — has to be regarded as the most difficult of his life. Only one very fond of the country can conceive how dreadful it was to BEIJERINCK to exchange rural Wageningen for a small factory town, devoid of all natural beauty, as Delft was in those days. Delft with its famous past, its old canals and buildings, with its mausoleum of WILLIAM THE SILENT and of the Kings of Holland, Delft of VAN LEEUWENHOEK, had become a declining provincial town.

His sisters remained in Wageningen, and therefore BEIJERINCK had to resort to a life in lodgings. Very soon after arriving in Delft he deeply regretted his decision and, since he never sought company, he led a life lonely in the extreme. He also regretted having thrown away his chance of a professorship at the Agricultural College at Wageningen for a career which in his opinion was difficult and full of uncertainties, and in which he was afraid of not being able to live up to expectations. He was subject to prolonged fits of depression, and his ever-sympathetic sisters often had need to encourage him. They came to see him frequently, visiting sometimes the laboratory. It may be interesting to quote from the diary of Miss H. W. BEIJERINCK. "He sits there surrounded by a number of retorts, bottles and glass boxes, gas ovens and heating apparatuses, so that it looks like the workshop of an alchemist. He is especially occupied with the investigation of bacteria which have an unfavourable influence on yeast cells and tries to cultivate the latter in such a way that they are free from bacteria".

Still there were bright spots, for, though BEIJERINCK did not get on well with his colleagues, he formed a close friendship with a young technologist, F. G. WALLER, a future Chairman of the Board of the Yeast and Spirit Works. This was a friendship which lasted till BEIJERINCK's death.

Afterwards BEIJERINCK used to tell that his first practical suggestion caused loss to the factory. He suggested to VAN MARKEN that the distillery slop should be used as food to pigs. VAN MARKEN appointed a veterinary specialist, and ordered a number of pigs, which greedily ate the stuff, but which for some reason developed black teeth, making them unmarketable. It was a good thing for BEIJERINCK that at that time Mr. WALLER was about equally unfortunate in his work for the factory.

Notwithstanding his discontentment, in December 1886 BEIJERINCK rejected the chance put before him of becoming director of one of the sugar experiment stations in Java. When some years later things at the factory had become very difficult, he was bitterly disappointed when nothing came of a post offered him on a sugar estate on Java, where he would have received an enormous salary.

Throughout his time at Delft BEIJERINCK was very well off and behaved very liberally in financial matters towards his family. He indulged himself in frequent holidays which he passed in foreign countries. He visited Switzerland several times. Little information reaches us about these jaunts, because BEIJERINCK nearly always travelled alone, and he never told much about his excursions, not even to his sisters [1]).

By 1890 he had come to feel so uncomfortable in his post, since he felt that he could not come up to the expectations people had had of him, that he spoke of resigning, hesitating to give his resignation more definitely.

This state of mind must undoubtedly be ascribed to BEIJERINCK's more or less neurasthenic proclivity, which sometimes made him place grave interpretations upon very innocent happenings. To his sisters he said that he was going to leave the works "unless a miracle took place". The sisters at once rented a house next-door to their own in the Dijkstraat at Wageningen, and furnished it for him.

However, the situation was — as many times before — saved by Mr. VAN MARKEN, who wrote BEIJERINCK a very tactful letter in which he was rebuked for his fickleness, but in which BEIJERINCK at the same time was assured that the Board of the factory indeed appreciated his work. So BEIJERINCK wired to his sisters: "The miracle has happened! They wish to keep me, and I wish to stay".

BEIJERINCK's troubles were also caused by his deep sense of failure in looking after the interests of the factory, the Board of which paid him so well and were so obliging to him. On studying his collected papers we see that besides researches on butyl alcohol fermentation, and *Schizosaccharomyces octosporus*, he also studied *Bacillus radicicola*, luminous bacteria, and green algae, subjects the relation of which with the technical trade is hard to find. The managing board showed themselves to be very broad-minded by allowing BEIJERINCK so much freedom in his scientific work and his publications.

Meanwhile BEIJERINCK was considered for the occupation of the chair of botany at the University of Groningen, as the successor of Professor DE BOER. Probably because BEIJERINCK asked for too much, the professorship was given to Dr. J. W. MOLL. From that time onwards the managing board of the factory seemed to have felt

[1]) He once remarked that at some time in Switzerland he had been wondering whether the diameters of the boulders at the feet of the glaciers would vary according to a "Galton"-curve.

that they were no longer justified in keeping BEIJERINCK confined within the factory buildings at Delft. In all probability it was owing to VAN MARKEN's influence with the government that attempts were made to procure him a position as professor of bacteriology. Plans were made, for example, to build a laboratory for agricultural bacteriology at Wageningen or at Utrecht, but already in 1892 it was decided to offer BEIJERINCK a professorship in Delft. The professor of chemistry, S. HOOGEWERFF, seems to have been mainly responsible for this decision. BEIJERINCK was greatly attracted by the idea of the latter, although his friend VAN 'T HOFF tried to persuade him he would do better to stay at the factory, alleging that he was too self-contained to become a good professor.

In the following years he made several journeys to foreign countries. One trip took him to London, for an investigation concerning the possible rôle of yeast as a carrier of cholera germs. At the end of 1892 he went to Paris to attend the celebration of the seventieth birthday of PASTEUR but, as was his habit, he said nothing about it to his sisters.

Next year he went to live with his sister Johanna at the Leeuwenhoeksingel in Delft, where he took a house; solitude had become too much for him.

In April 1893 BEIJERINCK entered into negotiations with the Director of the Polytechnical School at Delft, Professor A. C. OUDEMANS regarding the possibility of BEIJERINCK's professorship at this School. His main conditions were a new laboratory, and a salary of *f* 5400.—, which was extremely high for that time, higher even than that of the Director. All this was discussed in December in the House of Commons, on which occasion the Minister of Home Affairs promised to divide the salary into a normal fee and a personal gratification. There was a rather severe opposition to the proposal, but finally the motion was carried 42 to 36. *f* 20.000.— for a house, and *f* 45.000.— for the laboratory was voted. In February 1894 the plan for laboratory and house, to the design of Professor GUGEL, was passed. The building was to take place with the aid of a temporary superintendent, under the supervision of BEIJERINCK. A laboratory would be built with an upper part as living quarters. This plan, however, was rejected by the chief of the Government Architectural Department, VICTOR DE STUERS. In consequence of this the architects made a new plan for a laboratory and house adjoining in the Nieuwe Laan, everything being larger and better. BEIJERINCK left his post at the "Nederlandsche Gist- en Spiritusfabriek" on July 1st, 1895. On June 28th the news of his appointment as a professor was in the papers.

When this stage had been reached, the same happened to BEIJERINCK as at the time of his appointment as bacteriologist of the yeast factory. He wished to be quit of his new post, and he was very sorry to say good-bye to his comfortable life at the factory. Besides, he was afraid that he would not be able to command the attention of the students. He also feared that his lectures would not be well attended, because his subject was not compulsory for the examinations. Since he frequently suffered from mental fatigue, he thought that he would have to resign after a year. The telegram of congratulations from his sister Henriëtte he never answered at all!

The managing board of the Yeast and Spirit Works once more showed great liberality by placing at his disposal the laboratory of the factory during the time that the bacteriological laboratory was being built.

On September 26th, 1895, at the Polytechnical School, BEIJERINCK gave his inaugural address, entitled: "De biologische wetenschap en de bacteriologie" [1]).

---

[1]) "Biological Science and Bacteriology". Cf. Verzamelde Geschriften **3**, p. 154.

CHAPTER VI

THE ACADEMIC TEACHER

At the beginning of his career BEIJERINCK had to face several difficulties, including the envy of several of his colleagues that a newly-appointed professor should have a new laboratory, while they themselves had to work in old and cramped surroundings. The manner in which he had been appointed had also caused great annoyance. His lack of pliability was, besides, the cause of some friction with the Director of the Polytechnical School. In the years when the elevation of this School to a "Technische Hoogeschool" was being prepared, the Director said: "One thing must remain, and that is the Directorship of the school". BEIJERINCK answered: "Sir, if anything has to disappear, it is the Directorship!" Afterwards BEIJERINCK said to one of his friends about this: "The others were too cowardly to give me any support". He was also greatly annoyed that money was given for teaching purposes but never for scientific work; from this he drew the rather startling conclusion that the Minister of Home Affairs disliked him.

In April 1897 the house in the Nieuwe Laan was finished and BEIJERINCK with his sister moved from the Leeuwenhoeksingel to his new home adjoining the laboratory. Here the elder sister soon joined them. From this time on the trio remained united till death separated them.

On September 28th BEIJERINCK opened the laboratory by giving an address entitled "Het bacteriologisch laboratorium der Polytechnische School" [1]). The ceremony was attended by the authorities and several of BEIJERINCK's colleagues.

The inauguration of his academic career led soon afterwards to an event which rather characteristically typifies BEIJERINCK's mental state. On the first of October in that year the undergraduates serenaded him, as they always did with newly-nominated professors. A number of professors of the Polytechnical School with their wives were present at his house; among them being his friends, Professors ARONSTEIN, HOOGEWERFF, KREUNEN and PEKELHARING. Owing to the nervousness which usually overcame BEIJERINCK as soon as he had to act outside the scientific field, he made a speech to the undergraduates which in curtness and harshness could hardly be equalled. His au-

[1]) "The Bacteriological Laboratory of the Polytechnical School". Cf. Verzamelde Geschriften **3**, p. 233.

dience was so greatly taken aback that Professor HOOGEWERFF felt called upon to make amends by a more cordial speech. Then BEIJERINCK and his guests had supper, which ended in great exasperation to the host.

This is not the place to deal with the scientific activity displayed by BEIJERINCK and his collaborators during the twenty-four years he was in charge of the laboratory at the Nieuwe Laan. For a sketch of BEIJERINCK's scientific method the reader is referred to Chapter VII, whilst a detailed account of the more important investigations carried out during this period may be found in Part II and III of this biography.

It seems desirable, nevertheless, to say here something about BEIJERINCK's relations with his assistants and students.

BEIJERINCK was exceptionally lucky in the selection of his assistants, and this circumstance materially contributed to the success of his scientific work. Though it is impossible to mention all his assistants here — a complete list is given in Appendix B — a few words may be devoted to some of the more prominent amongst them. Plate VI presents their contemporary portraits.

BEIJERINCK seems never to have lacked an appreciation of the importance of salaries, and he succeeded in obtaining for his assistants a remuneration considerably higher than the normal. Thus, his assistants had no immediate reason to be on the look-out for better-paid jobs, and several of them remained in office for relatively long periods.

BEIJERINCK started his work in 1895 with only one assistant, A. H. VAN DELDEN, a young technologist who had taken his degree only shortly before. Although VAN DELDEN entirely lacked experience in microbiology, he soon developed into a very able bacteriologist. VAN DELDEN stayed with BEIJERINCK until 1904, when he accepted the post of bacteriologist of the Rotterdam Water Works, of which he later became an assistant director [1]). It is certain that the period of VAN DELDEN's assistantship covers that of BEIJERINCK's greatest achievements in the microbiological field. It is difficult to estimate correctly the part which VAN DELDEN had in many of BEIJERINCK's investigations, but there is good reason to suppose that VAN DELDEN's share was far from negligible. BEIJERINCK did not always stop to consider the justice of giving credit where credit was due in the publication of results of joint work. The fact that BEIJERINCK's strongly-marked individuality ceded to VAN DELDEN the right to join his name to BEIJERINCK's in papers on nitrogen fixation, on *Bacillus oligocarbophilus*, and on the retting of flax, leaves no doubt that VAN DELDEN's contributions to these studies must have been substantial. VAN DELDEN published separately an important paper on sulphate

[1]) VAN DELDEN died in 1926, at the comparatively early age of 52.

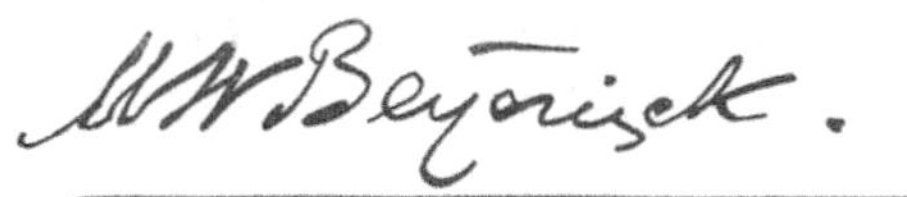

Beijerinck in the prime of his life, at the age of 45.

A. H. van Delden.

G. van Iterson Jr.

H. C. Jacobsen.

N. L. Söhngen.

Four prominent collaborators of Beijerinck during his academic period.

reduction, following up BEIJERINCK's earlier investigations, in which he had assisted (Cf. Appendix C).

VAN DELDEN was a very modest and unselfish person, and was devoted to the man who had done so much to widen his scientific horizon.

In 1902 BEIJERINCK obtained a second assistant on his staff. He was, moreover, so fortunate as to find a very competent candidate. Struck by the exceptionally fine way in which G. VAN ITERSON Jr. had taken his final degree, he invited this young scientist to become his collaborator. After some hesitation, VAN ITERSON — who until that time had been specializing in physical chemistry — accepted the invitation. There is no doubt that VAN ITERSON is by far the most brilliant pupil BEIJERINCK ever had. VAN ITERSON quickly exhibited great activity, and his independence being apparently a match for BEIJERINCK's, he laid down the results of his investigations in several publications under his own name (Cf. Appendix C). His studies on denitrification and on the aerobic decomposition of cellulose have proved to be of a fundamental nature. Gradually his interest shifted more and more to the field of general botany. His Doctorate thesis, entitled "Mathematische und mikroskopisch-anatomische Studien über Blattstellungen", bears witness to his remarkable achievements in this field. That BEIJERINCK had a great admiration for the scientific capacities of his collaborator may be judged from the way in which he once introduced VAN ITERSON to the then Minister of Home Affairs, Dr. A. KUYPER, who paid a visit to his laboratory. BEIJERINCK said on this occasion: "This is Mr. VAN ITERSON, my assistant, who knows much more than I do".

VAN ITERSON's scientific evolution soon made it clear to BEIJERINCK that his assistant was the right man to accept responsibility for part of the teaching. To begin with, he made VAN ITERSON organize a special course on plant anatomy, but it soon became apparent that this part of the curriculum of the chemistry students would be able to flourish only, if more material support could be provided. Therefore, shortly after VAN ITERSON had taken his Doctor's degree, a new chair of "technical botany" was created for him, and he was thereupon moved to a new laboratory especially equipped for the study of pure and applied botany.

A third assistant, one whose activities have undoubtedly been of great significance for the development of BEIJERINCK's work, is H. C. JACOBSEN. He succeeded VAN DELDEN in 1904, and holds the record for length of service, for he did not leave the laboratory until 1916. He then became bacteriologist to the Jurgens Margarine Works, later amalgamated into the Unilever concern.

The articles which JACOBSEN published in his Delft period under his own name can be found in Appendix C. Amongst them, his investigations on the unicellular alga *Haematococcus pluvialis*, and on various *Volvacaceae*, deserve a special mention. In addition JACOBSEN most unselfishly did an enormous amount of work to support BEIJERINCK's researches during the second half of his academic career. Moreover, he considerably lightened BEIJERINCK's task by taking over a part of the instruction of the less advanced students. The favourable influence JACOBSEN had on the course of affairs at the laboratory in the Nieuwe Laan cannot easily be overrated.

Also N. L. SÖHNGEN largely contributed to the scientific standing of the institute, as appears from the numerous articles published by him during his stay at the Delft Laboratory (Cf. Appendix C). SÖHNGEN was the first to take a Doctor's degree at Delft, after the new Higher Education Act made that possible by bringing about the conversion of the Polytechnical School into a "Technische Hoogeschool". SÖHNGEN's thesis dealing with the production and consumption of methane and hydrogen in nature has now generally been recognized as a classic. Yet it seems that at first BEIJERINCK did not feel much inclined to accept this thesis as such; apparently he shrank from the troubles involved.

Soon after he had obtained his doctorate, SÖHNGEN left Delft and acted as bacteriologist to some margarine factories in Rotterdam and in Middelburg. In this period he published several papers, some on the bacterial decomposition of fats, and others on urea fermentation. On December 1st, 1911, however, he accepted the post of assistant at Delft and held that post until September 1915. In this second Delft period he studied amongst other subjects the mineralization of hydrocarbons like benzene, kerosene, etc. He also published an extensive study dealing with the factors causing offensive odours in the canals at The Hague.

SÖHNGEN's independent character prevented him from cooperating closely with BEIJERINCK. In 1915 SÖHNGEN became Director of the Microbiological Division of the Government Agricultural Experiment Station at Groningen. In 1917 he was appointed professor of microbiology at the Agricultural College at Wageningen, where he remained until his death in 1934. Over this period he did a great deal to propagate the application of BEIJERINCK's science to agricultural problems.

We have no space to mention the work of the other assistants who were for the most part temporary.

The frequently impossible demands BEIJERINCK made on his assistants often caused somewhat strained relations between them and him. It was no light task to be his demonstrator. The junior might do his best, but was often grumbled at by the professor just before the lecture. At such moments BEIJERINCK was always more or less nervous, and often managed to set his demonstrator on edge as well. Sometimes the poor fellow was the target for a sneer during the lecture for some "carelessness" or other, and after the lecture, when the experiments for next time were discussed, his sins were sometimes brought up again! It happened on occasion that BEIJERINCK arranged a social evening for students at his house but forgot to invite the assistants. When, afterwards, he tried to make amends for his negligence his genuine remorse was almost painful to witness.

It was a matter of keen regret to BEIJERINCK that especially in the later years only a comparatively small number of students attended his lectures. This was no doubt due to the circumstance that the study of microbiology in Delft was not compulsory. Nevertheless it is indisputable that BEIJERINCK put his stamp upon the scientific development of those students who worked for any considerable period in his laboratory.

The number of students who stayed with BEIJERINCK long enough to take a Doctor's degree under his direction was, however, not large. Appendix D gives a list of their theses. It must be added that several of them were the result of experimental work partly or entirely done elsewhere (VAN HALL, RANT, HEYMANN, and GERRETSEN).

The initiation undergone by students in BEIJERINCK's laboratory was too searching to be pleasant. They were weighed and often found wanting, and woe to them when this was the case! So little a thing as a drop of water spilt on the bench — which drop was then demonstratively removed — might give rise to a burst of anger. Not only had the students got to listen again and again to a summing up of all the stupid things they had said or done, but also they were told of all the blunders they were likely to make in future.

Characteristic of BEIJERINCK's attitude of mind towards his students is the following speech made to a victim who had failed to give a correct answer to one of BEIJERINCK's questions: "Sir, there are two types of monkeys. One type is interested if one shows a coin and will hold it firmly, the other type will at once drop it. The first type can be trained, the second type cannot. If you were a monkey, you would belong to the second type!"

A good idea of the atmosphere in BEIJERINCK's laboratory was given by Professor JAN SMIT in his obituary speech [1]) entitled "BEIJERINCK's levenswerk" ("BEIJERINCK's life-work"), here translated:

"Then began a period of restless scientific work with the co-operation of a great number of pupils from Holland and abroad. It is almost impossible to give an adequate idea of it. One has to have witnessed the high tension found there, and to have heard the conversations, sometimes lasting for hours, with one of the experimenters, where usually BEIJERINCK did the talking and the other the listening — fascinated by the stream of surprising and new remarks with thousands of suggestions for new experiments which the professor poured out over his unfortunate head. The student tried to take it all in, but at last was almost in despair, because his head was unable to contain that overpowering amount.... while BEIJERINCK, as fresh as

[1]) Chemisch Weekblad **28**, 94, 1931.

if nothing had happened, went on to another student to lose himself entirely in the latter's subject. And every student could be quite sure that at a following visit the professor would inquire into the progress made, and would not hide his displeasure, should any one of the many experiments recommended by him not have been made. BEIJERINCK was not easy to deal with. He did not ask less of those who worked with and for him than of himself. He would dash through the laboratory like a whirlwind, shutting all the windows on the way, with never-failing accuracy immediately detecting any clandestine cigarette smoke, and withering its originator with a look as if the cigarette were a venomous insect! And woe to him who during the daily conversations betrayed lack of care in studying his subject, or indulged in experiments which were unimportant or did not apply to the subject in hand! Such a "bungler" was only left the choice between an immediate return to the right track or complete self-contempt! But whoever came to him with the fixed purpose of learning as much as possible found an inexhaustible source of knowledge from which he might drink, even to suffocation".

BEIJERINCK was always ready to help his students by word and deed, and a number of them owe to him a prosperous career. It has happened that he stood up for a student who had incurred the displeasure of the other professors, and helped him on again. He was very compliant to his former students, and assisted them as much as possible in all cases, whenever they applied to him [1]).

In spite of all this, BEIJERINCK was never popular as a professor. We need not look far for the reason. He was that paradox, the personification of impersonal science. His whole personality had been absorbed in it. Other things practically did not exist for him. His life as professor was that of a recluse, as in the preceding period, although it has to be remarked that he was a very regular attendant at meetings where duty called him, such as those of the Faculty. Social gatherings, dinners, etc. were his abominations, to be avoided as much as possible. They always gave him a bad headache.

It almost stands to reason that BEIJERINCK never got married. Once he had a great disappointment. He did not always approve of marriage in his collaborators either. Very characteristic in this respect was his answer to the announcement of the marriage of one of his assistants: "A man of science does not marry". Once BEIJERINCK happened to witness a harmless flirtation between a boy and a girl student in his laboratory. The explosion of anger which followed this innocent event surpassed all rational limits. Such a behaviour he considered to be a profanation of his laboratory, and of science in

---

[1]) A typical illustration of BEIJERINCK's spontaneity and helpfulness is the fact that, when in 1920 fire broke out in the house of his faithful amanuensis KOKEE, BEIJERINCK offered him and his family hospitality in his laboratory, where they lived for several months.

general! After this it will hardly be necessary to say that the girl-students were not in his good graces. His lectures were always opened with "Gentlemen and Ladies!"

Clearly he was never *en rapport* with young people — such as his students — whilst on their part the majority of the students regarded him as the most crusted example of a "professor" that could possibly be conceived.

It is not doubtful that this situation was chiefly due to the fact that BEIJERINCK belonged to those persons who seek and love solitariness, needing it to think out their thoughts, and to assimilate their impressions and experiences. He needed solitude also because he had to interchange periods of great physical and mental stress with moments of rest and restriction of mental activity.

In the beginning of his career as a professor BEIJERINCK apparently did not suffer at all from this lack of human contact. Every impression awoke in him so many recollections of earlier experiences, and stirred him up to so numerous critical reflections that any feelings of loneliness were soon repelled.

This may also suggest an explanation why BEIJERINCK so rarely kept himself between bounds in his intercourse with collaborators and students.

Even if he had a personal appreciation for the man in question, he often sallied out in a way which was not at all justified. In these earlier years he placed no value on friendly relations with his collaborators, and he was quite content for their feelings to be restricted to nothing warmer than admiration and astonishment.

A factor in the formation of this detached attitude was possibly BEIJERINCK's gradual perception — based on unfortunate experiences — that contact with other persons might give rise to conflict. For the sake of peace, therefore, he sought only to be in contact with people possessing an unrestricted admiration for him, or with those who had unfailing patience and the power to forgive and to forget.

With increasing age BEIJERINCK lost something of his egotism. A disclosure from Miss BEIJERINCK's diary illuminates the man as he was in later years. The various disagreements BEIJERINCK had with his assistants, the small size of his classes, and the lack of warmth displayed towards him by the students who did come, all this worried BEIJERINCK far more than anybody could have thought. Deep down in his heart he needed sympathy and kindness, and he did not receive either, because hardly anybody suspected him of needing it. To the outer world he was the fossilized scholar, a stranger to human feelings.

Yet, it should not be concluded from this that he was not appreciated. Once his peculiarities were forgotten, irritation gave way to reverence. BEIJERINCK was like a mighty building. Wandering through its unfamiliar courts and archways, a visitor might sometimes knock against, and be hurt by, protruding stones, but after leaving the building and contemplating at a distance its superb architecture, the former visitor would be lost in rapture.

The description of BEIJERINCK's activities in his academic period would be incomplete, if no mention were made of the fact that a quite considerable number of foreign scholars came to work some time under BEIJERINCK's supervision. Amongst these there were several scientists who have gained a well-deserved reputation in their special fields. The names of ISSATCHENKO (Russia), GRAN (Norway), KASERER (Austria), STOKLASA (Austria), KRZEMINIEWSKI (Poland), KRAINSKY (Ukraine) and MELIN (Sweden) may be recorded here.

The professorship often weighed heavily upon BEIJERINCK. Already in 1905, following a rather serious difference of opinion with one of his collaborators, he felt inclined to resign. This urge came with new force a few years later, when many of his colleagues, such as Professors HOOGEWERFF, ARONSTEIN and SCHELTEMA, had left. In 1912 the intention turned up again, but, when after great trouble the laboratory at the Nieuwe Laan acquired a new annexe, he was again able to enjoy the work, although it was still disappointing, after all the material improvements, that only a few students came to work with him.

There is no doubt that in the second half of his professorship BEIJERINCK was often dissatisfied with his achievements. For instance, he once exclaimed: "At Delft I have come to grief! If I had remained at Wageningen, I should have been ahead of BUCHNER with his discovery of the press-juice, I should also have rediscovered the Mendelian laws, and those are not the only things I have missed!" He apparently often dreaded a decline in public recognition, and he also fancied that he no longer came up to the demands of his position. Indeed there is reason to think that by 1905 he had attained his scientific peak. This may have partly been due to the circumstance that exploration of the microbe world, which under BEIJERINCK's pioneering guidance had led to so many remarkable discoveries, had by then entered on a new and more settled phase.

Other matters certainly contributed to BEIJERINCK's periodic attacks of mental depression. In 1911 the early death of his friend VAN 'T HOFF, then professor at the University of Berlin, made a deep impression on him. He at once went to Berlin, and also attended the cremation at Ohlsdorff near Hamburg.

Great was his distress when war broke out in 1914 and almost entirely put a stop to his relations with foreign scientists.

At various times he intimated that he feared approaching old age. But, in spite of all his worries, BEIJERINCK did know many moments of real happiness. Possessed of an urge after knowledge, an almost Dionysian joy often came over him when his experiments were successful. Then his brown eyes would glitter, and, with a staring look and uplifted left forefinger, he would explain the significance of his discoveries to his disciples. In doing so he often railed at the many mistakes made by his predecessors in studying the subject in question [1]).

It should not be concluded from this that BEIJERINCK never showed any respect for the work of other investigators. As a matter of fact he had a profound admiration for the great founders of science, especially for the biologists, such as van LEEUWENHOEK, PASTEUR, and DARWIN. But he also worshipped physicists like NEWTON and FARADAY, the first of whom he, however, could not forgive for having spent so much time on the exegesis of the Apocalypse.

BEIJERINCK's activities as a professor were marked also by a long-sustained interest in general botany. Not seldom he passed a considerable time in his garden, in which many rare species were assembled, and in which BEIJERINCK often made remarkable experiments. Sometimes his enthusiasm for the results obtained was so great that he commanded his students to join him in the inspection.

It is noteworthy that BEIJERINCK became deeply interested in problems connected with the squaring of the circle, trying to relate these with phyllotaxis. The latter phenomenon occupied him till the last days of his life, but he never came to a conclusive treatment.

Apart from his purely scientific work, BEIJERINCK gave attention to problems in the field of applied science.

For instance he acted, as an adviser to the flax industry, for which his laboratory studies on the retting of flax had pre-eminently qualified him. Moreover, he was adviser to the "Nederlandsche Heide-Maatschappij" [2]), a member of the Board of the State Agricultural Experiment Stations, and of the State Committee for the Purification of Sewage; an adviser of the State Institute for Fishery Research, and during the war he was a member of the State Committee for Public Welfare and Defence.

It is not surprising that BEIJERINCK's many-sided activities brought him several marks of respect and recognition. The government acknowledged his merits by making him as early as 1903 a Knight of the Order of the Nederlandsche Leeuw, and at his resignation in 1921, a Commander of the Order of Oranje-Nassau.

That the great significance of his scientific work was already early recognized by his fellow-countrymen appears from the fact that in 1898 he was offered a professorship in botany at the University of Leiden, in succession to his former teacher, Professor SURINGAR. BEIJERINCK, however, declined this invitation.

The distinction which BEIJERINCK valued most was the award, by the Royal Academy of Sciences at Amsterdam in 1905, of the LEEU-

[1]) A favourite expression in this connection was: "een echte vieze knoeier" (a real dirty bungler).
[2]) "Netherlands Society for Heath Reclamation".

WENHOEK Medal. BEIJERINCK was always deeply impressed by the genius of this great naturalist.

The medal was conferred upon BEIJERINCK in the meeting of the Academy of September 30th, 1905; Professor F. A. F. C. WENT, the later well-known botanist, gave an address that testified to his profound admiration for BEIJERINCK's work. BEIJERINCK made a brief reply. Both speeches have been reprinted in Appendix E.

In 1906 BEIJERINCK was made an honorary member of the Royal Botanical Society of Edinburgh. In 1917 Professor ORLA-JENSEN of Copenhagen informed him that he had proposed him as a candidate for the NOBEL prize for chemistry. However, nothing came of this; in that year the prize was not awarded. He also received invitations from Berlin (through VAN 'T HOFF) and from America to lecture: invitations, which however, he did not accept.

H.M. the Queen and H.R.H. Prince HENDRIK OF THE NETHERLANDS, and such highly-placed persons as the Prime Ministers GOEMAN BORGESIUS and ABRAHAM KUYPER gave evidence of their interest by paying visits to his laboratory, marks of honour such as have only rarely been accorded to a Netherlands man of science.

After the war was over, BEIJERINCK again began to think seriously of resigning his professorship, but finally he decided to remain till his 70th year, *i.e.*, till 1921.

His 25 years' professorship on July 1st, 1920 passed almost entirely unobserved. BEIJERINCK and his sisters escaped from Delft; but his friend and ex-colleague HOOGEWERFF wrote a commemorative article which was published in the weekly periodical "De Ingenieur" (Cf. Appendix F).

In contrast to this, a grand celebration took place about a year later on the occasion of his seventieth birthday, *i.e.*, on March 16th, 1921. Months earlier, a committee had been formed of friends and ex-pupils to organize this celebration. On the eve of the big day BEIJERINCK himself, however, seems to have been too much impressed by the forth coming end of his academic career to enjoy the prospects of the homage which was to be paid to him. A proposal made by his staff to hoist the national flag was rather ungraciously dismissed with the words: "One does not hoist flags on the day of one's funeral". Notwithstanding this, there is no doubt that at the end of the day, BEIJERINCK was extremely gratified by all the honour done to him.

As a matter of fact, the committee, and especially its president, Professor VAN ITERSON, had spared no pains to make this day unforgettable for BEIJERINCK. Large funds had been raised to which his friends and more distant admirers had contributed; the Netherlands Government had also made a considerable contribution. Part of the money thus raised had been used to finance the publication of BEIJE-

RINCK's "Collected Papers" in five large and dignified volumes [1]). A considerable sum remained over. This was handed to BEIJERINCK in order to enable him to build a private laboratory after his retirement.

At the celebration itself the first volume of the "Collected Papers" was presented to BEIJERINCK by Professor VAN ITERSON, who had previously made an eloquent address surveying BEIJERINCK's scientific work. This address has been reprinted in Appendix G. His old friends Dr. F. G. WALLER and Professor HOOGEWERFF made shorter speeches, in which they testified to their great sympathy and admiration. Professor KRAUS, President of the Board of Curators, handed BEIJERINCK the cross of Commander of the Order of Oranje-Nassau.

On April 21st, 1921 BEIJERINCK was relieved from his office to date from September 6th, with thanks for the many and important services rendered to the country, and on May 28th the farewell lecture was given in the lecture hall of the chemistry laboratory.

Characteristic for BEIJERINCK's scientific attitude is that he wished to enliven also his academic swan-song by inviting his assistant — the present author — to make demonstrations of various microbial cultures. The title of the lecture, an abstract of which is given in Appendix H was: "De cel; erfelijkheid en variabiliteit bij de microben" [2]). A great number of colleagues, students and friends attended the lecture.

BEIJERINCK's concluding words are especially noteworthy:

"When a leaf drops from the tree, it happens because a partition-layer of live cell-tissue has been formed between branch and leaf. At the moment of dropping, the partition layer is split in two, by a pressure developing which disconnects the bundles of vessels, *i.e.*, the threads of communication between branch and leaf. One half of the partition layer remains on the branch, the other on the leaf. The tree is the Technische Hoogeschool, and the branch the department, the dropping leaf is the parting professor, the pressure causing separation is the law. The twinned partition layer is the remembrance. This will last for some time on either side; on the branch, in the department, until growth shall obliterate the last traces. This will be for a long time for those who come after us will find our names in the records of the Technische Hoogeschool, and will ask who we were. But the leaf with its share of the partition layer will soon decay, as will the departing professor, who takes his memories with him till the moment when he himself ceases to exist...."

BEIJERINCK dutifully stopped in Delft until the examinations were over. He did this chiefly because his two latest assistants had to take

[1]) The sixth and final volume appears simultaneously with this biography.
[2]) "The cell; heredity and variability in microbes".

their final degree. It was not until June 21st that he and his sisters left Delft for his summer residence at Gorssel. The hour of departure was undoubtedly one of the saddest in BEIJERINCK's life. In the quarter of a century which he spent at the laboratory in the Nieuwe Laan he had become so enmeshed in his surroundings that he came to look upon them as his private property. The necessity of departure he felt as an act of injustice. This feeling may have contributed to BEIJERINCK's never returning in Delft.

CHAPTER VII

BEIJERINCK AT WORK [1]

Before proceeding to an account of the last period of BEIJERINCK's life, it seems fitting to give here a brief analysis of BEIJERINCK's methods of working. It is not intended to enter into details regarding BEIJERINCK's approach to the numerous problems which occupied his attention. Such matters will be adequately dealt with in Parts II and III of this book. However, it is felt that a description of BEIJERINCK's general laboratory technique, and a consideration of the factors which determined the choice of the subjects of his investigations may not be omitted here.

Above all, it should be stressed that BEIJERINCK was an exceptionally keen observer. The starting point is for most of his investigations, especially in the first phase of his scientific activity, to be found in observations made either in nature, or during laboratory experiments. His classical studies on galls and gall insects are characteristic of his rare gifts for observing plant and insect life, and for giving an interpretation of the phenomena observed.

However, BEIJERINCK was also a man of great learning and wide reading, and, during his later years especially, it was often something he had read in scientific literature that stimulated him to undertake investigations which frequently led to new and unexpected results.

BEIJERINCK's ability to fuse the results of remarkable observations with a profound and extensive knowledge of biology and the underlying sciences has undoubtedly been responsible for the great success of his work.

Characteristic of BEIJERINCK's mode of attack of the various problems is the fact that he generally started with a quite definite "working hypothesis" which was then submitted to well chosen experimental tests. Depending on the results obtained the hypothesis was amended, and then, once more, critically tested. In these experiments BEIJERINCK profited by his ever increasing experience, and — unlike more specialized investigators — he was able to mobilize his great empirical knowledge of the fields of physiology, experimental morphology, chemistry and physics. However, the applications he

[1] In the drawing-up of this chapter the author has freely used, and has greatly profited from an exposé by Prof. VAN ITERSON, who during many years, in which BEIJERINCK's scientific activity culminated, was in daily contact with the great scientist. He also wishes to gratefully acknowledge several contributions made by Mr. H. C. JACOBSEN, for whom the same holds.

made of the latter fundamental sciences were nearly always restricted to those which he could easily verify in his own experiments; phenomena which lay beyond that boundary did not interest him. Even his knowledge of mathematics was mainly of an experimental character, he deduced mathematical theses by geometrical constructions. Especially in the latter part of his life he was not seldom led astray by this procedure, he then "discovered" peculiarities which could have been proved — or disproved — in a much simpler way.

Another feature of BEIJERINCK's life work is its great diversity. The contents of his "Collected Papers" reveal an astonishing variation in subjects, even of consecutive papers. BEIJERINCK's mind was so rich that he could not centre his attention for long on any one subject. New thoughts continuously took hold of him and forced him to leave recently-conquered ground. In consequence of this, as a rule, BEIJERINCK was occupied with several problems at the same time. Even in the period of his microbiological activity he not seldom returned to purely botanical studies.

This mercuriality of BEIJERINCK's mind was also clearly manifest in his conversations, and — at least in later years — in his lectures which, although always sparkling, often lacked logical consistency.

All this should not be understood to mean that BEIJERINCK had no general scientific program. On the contrary, the chief aim in his microbiological work has always clearly been to create order in the chaos of the microbe world. The very consciously-applied "enrichment culture method" led to the recognition of numerous well-defined physiological groups in the bacterial kingdom, and has furnished a stable basis for future work.

Nor can it be said that BEIJERINCK was always unfaithful to those subjects which had once engaged his full interest; he often came back to an old theme after several years, and then dealt with it in a broader and more profound way.

The scientific unrest which was so prominent in BEIJERINCK on the other hand explains why he was never able to persuade himself to write textbooks or monographs. Such writing demands introspection and patience.

The most marked trait of BEIJERINCK's scientific personality was undoubtedly his passion for experiment. However, he liked only simple, and if possible, elegant experiments. Galvanometers, registration apparatuses, etc. are nowhere mentioned in his papers. One could almost say that he abhorred complicated instruments, and one of his favourite sayings was: "An experiment should be simple". This attitude was also manifest in the use of the microscope. Oil immersions were rejected as being "dirty", dark field illumination, as being too complicated. His usual optical outfit consisted of a ZEISS microscope with achromatic objectives 8 × and 40 ×, and the from the optical standpoint rather unsatisfactory dry system 90 × (N.A. =

Beijerinck shortly before his retirement from the chair at Delft, at the age of 70.

0.90). The latter was mostly used with an ocular (5 ×), so that the magnification of the various micro-organisms was restricted to 450 times. Moreover, in his microscopical work BEIJERINCK often violated various optical rules; he never took into account the thickness of the coverglass of his preparations, the tube-length was never adjusted at 160 mm etc. Notwithstanding all this, BEIJERINCK — like the famous founder of microbiology, ANTONY VAN LEEUWENHOEK — owing to his excellent eyes and his keen observational power, generally saw more and better than the average microscopist. If BEIJERINCK now and then invited his assistants or students to admire his microscopical preparations, they often had great difficulty in checking his observations. In this connection it is worth mentioning that BEIJERINCK hardly ever made stained preparations; he considered staining an objectionable habit of medical bacteriologists, leading to the production of artefacts which veiled the real situation.

Photomicrography was always left to his assistants, but this did not mean that BEIJERINCK would not severely criticize the results!

Although he avoided all complicated constructions, BEIJERINCK was keen on designing simple and handy instruments. Mention may be made of his culture dishes with flat-ground covers which he greatly preferred to ordinary PETRI dishes. A further example is afforded by the special device he invented for the cultivation of anaerobic microbes in the presence of *Oidium lactis,* which organism was applied to remove the last traces of oxygen. Another of his small inventions was launched under the barbaric German name of "Kapillarhebermikroskopirtropfenflasche".

BEIJERINCK used to complain about his defective chemical education, but he exaggerated badly in saying that he knew practically nothing of this science. His fine investigations on the action of various enzymes, on bacterial pigments, and on the chemical constitution of microbial cell walls, and especially his work on sulphate reduction, denitrification and nitrogen fixation, testify to the contrary. There is ample evidence that next to biology, chemistry was his great love, and also that he had a deep insight into the chemistry of living cells.

In his own chemical work BEIJERINCK was especially fond of applying all kinds of qualitative analytical tests.

Quantitative estimations were as a rule too cumbersome for him; he left these to his collaborators. His appreciation for "weight and measure" was rather restricted, and in those cases in which he was inclined to accept a quantitative standard, he was easily satisfied with an approximate result. One of his favourite expressions was the paradoxical phrase: "I have investigated this somewhat quantitatively"!

His essentially "qualitative" work was, however, not seldom characterized by its elegance and its aesthetic qualities. We need here recall only his auxanographic method, his experiments on "micro-aerophily", and his remarkable demonstrations regarding the properties of luminous bacteria, over which his audience sometimes went into ecstasies. The artistic thread in BEIJERINCK's mind is also manifest in his often exceptionally fine drawings. The plates he added to his Doctorate thesis on the galls, the drawings in his papers on "Root-buds and adventitious roots", on the gummosis of the *Amygdalaceae,* on the green algae, on *Bacillus cyaneofuscus* etc., all give proof of this. It should not be forgotten, however, that his sister Henriëtte sometimes very ably assisted him in this part of his work.

The aesthetic element in BEIJERINCK's character may perhaps also explain why he was apparently especially attracted by those microbes which display beautiful colours in their cultures. We refer here to his studies on *Bacillus prodigiosus, Bacillus violaceus, Bacillus cyaneofuscus,* the "litmus-*Micrococcus*", chromogenous yeasts, green and blue algae, etc.

The beautiful blue-green sheen of the cultures of luminous bacteria may be at least partly responsible for the persistency with which BEIJERINCK studied these organisms. Colour-reactions, such as the cholera-red-reaction, the tyrosinase-reaction, indigo-formation etc., also occur frequently amongst BEIJERINCK's subjects of investigation. A predilection for regular structures such as can be for instance observed in microscopical preparations of *Schizosaccharomyces octosporus* and of *Sarcina ventriculi* may have contributed to the great attention he repeatedly gave to these organisms.

In the same line of thought BEIJERINCK's sense of smell may have led to studies on organisms producing attractive or in other ways characteristic odours, as is for instance the case with the acetic ester yeasts.

BEIJERINCK's working environment was generally characterized by a picturesque disorder. Although he easily became angry over similarly untidy work amongst his pupils, he was as a rule blind to his own shortcomings in that respect. His microscope was usually surrounded by dozens of inoculated culture flasks, numerous piles of culture dishes, bottles with reagents, and "BEIJERINCK-shelves" loaded with tubes containing pure cultures, so that he barely had room to move his arm.

In the second phase of his scientific career the plate culture was his proper field of operations. This field was explored in a manner that has probably never since been equalled. BEIJERINCK used to emphasize that a careful and circumstantial inspection of a plate culture was an unfailing source of knowledge, and it was only with heartfelt grief that he finally parted with the often already quite dried-up plates. It is difficult to estimate the pains taken by his patient "amanuensis" KOKEE in the preparation of the endless series of culture media. Every morning this worthy collaborator was summoned by a press of the bell-button, and then in a lengthy conversation hundreds of instructions were given for the correct preparation of the various media. Often very unusual procedures had to be applied; it is said that the only complaint which ever passed KOKEE's lips was: "How exacting the Chief was again this morning"!

The surface of the freshly prepared agar-plates was always dried by BEIJERINCK himself. For this operation the lower side of the dish was gently heated with a small gas flame, and the dish then deposited on the bench. The water vapour gradually condensed on the lower side of the cover which was then removed and dried with a clean handkerchief! For the sake of sterilization the cover was flamed, and again put on the dish. The plates, thus prepared, the culture flasks, and the tubes, were then inoculated — in later years with a trembling hand — in a way expressive of devotion, as if it were a religious act. The inoculated media were finally put in one or more incubators to which — of course — no other worker had access.

The next day, or the day after, the cultures were examined, were smelt, and — if possible — were even fingered. In special cases BEIJERINCK went so far as tasting some of the cultures!

Then the cultures were carefully examined, first with a pocket-lens with low magnification, next by putting the plates themselves under the microscope. Often the great scholar was seen sitting, till far into the night, bent over his microscope, delighting in the creeping of soil amoebae over the surface of the plates, or piercing with his eyes into the virgin forest of some fungal colony.

The solid media were often powdered with various compounds, in order to obtain special growth effects, increase in luminescence, etc. In other cases, drops of dilute acid or of alkali were put on the surface of the plates, and the effect thereof observed. Small pieces of the plates were cut out, and subjected to various treatments.

In a word, everything that could possibly be done with a culture plate, was done with it!

Only after the inspection of the plate was quite finished were the individual microbes studied in microscopical preparations, and one could be certain that here too, no detail escaped BEIJERINCK's attention.

It will be clear that such a procedure was extremely propitious for the discovery of microbial variation; and the corresponding inspection of liquid culture media was favourable for a minute analysis of fermentation, agglutination and similar phenomena.

All these observations unchained in BEIJERINCK's mind a wild current of thoughts, and he then would often give free play to his fancy. In such a mood he often called for his assistents, who then were regaled with an elaborate discourse over his bacteria; the bacteria almost were raised to the rank of human beings, as may be judged from a pronouncement like: "You could not have done, what I succeeded in doing this morning, for they (*i.e.*, the bacteria) don't know you!"

Fancy unbridled sometimes made him neglect well-established facts, and on occasions brought him into more or less dangerous situations. Although BEIJERINCK himself was aware of this, he would at such a parlous moment defend his attitude by a remark like the following: "One should dare to enunciate an idea, although it still remains unproven: another investigator can then refute the theory. This is the way in which science advances." Fortunately this perilous tendency was kept in check as a rule by his respect for the experiment.

It goes without saying that the way in which BEIJERINCK discussed the various phenomena he observed lent a special lustre to them, and that thereby he greatly stimulated his pupils and co-workers.

Finally a few words may be devoted to the way in which BEIJERINCK wrote down the results of his investigations. This operation proceeded only slowly and with great difficulty, probably owing to the whirlpool of thoughts which continuously took hold of him. He was never satisfied with the result, the text would be repeatedly rewritten, and after the unfortunate compositor had at last deciphered the pothooks and hangers of the manuscript, he would be sure to receive the proofs in a badly battered state. In one case, BEIJERINCK was heard to say after receiving the second proof: "Now is the time to interchange the columns in the tables"!

Beijerinck's country-home at Gorssel
(after a water-colour by his sister, Miss H. W. Beijerinck)

CHAPTER VIII

THE RETIRED SCHOLAR

Since 1902 BEIJERINCK has owned a plot of ground, on which he had built a cottage, and he was in the habit of spending the summer holidays there with his nearest relatives. Now he settled in a more comfortable house next-door to it, with a large garden adjoining. An idea of the idyllic situation of this last home of the great scientist is given by the water-colour painted by his sister Henriëtte, reproduced in Plate VIII.

In his new surroundings BEIJERINCK was able once more to devote himself entirely to botany; soon the place was transformed into a wonder-garden full of botanical curiosities, where he showed his visitors round with great enthusiasm, and was never tired of explaining everything in detail. The gardeners here — like those at Delft — had to steer a difficult middle course to satisfy their irascible master. If they did not keep the garden in order, they were stormed at, but if they cleared away "too much", *e.g.*, by hoeing away a particular weed in the middle of the path, they were rated still more: they should have had the sense to see and understand that this little plant of all others should have been spared....!

In the first years of this last period of his life, it was a delight to spend a week-end at the "Kleine Haar", as the country-seat was called. In Plate IX one sees BEIJERINCK as he was in these happy years, in an informal photograph taken by the distinguished microbiologist Professor S. A. WAKSMAN, when he visited BEIJERINCK in 1924.

When the visitor got out of the 'bus which stopped right in front of the house, the great scholar came to meet him with outstretched hand, asked in a friendly way how he was, took his case out his hand and led the way to a tree where he had just discovered a remarkable beetle. There they entered into an academic discourse which might last for half an hour, till BEIJERINCK came to the conclusion that the tired traveller might wish to refresh himself, and took his case to the guest-room, then waiting for him impatiently in his study. Here an enormous discharge of ideas and opinions came out in spate, and, the old Delft habits still being strong, the guest was taken to task for his ignorance, the lecture being followed up by a brilliant exposition of the right answer as it should have been given. Then a walk through the woods of Gorssel, which might last for hours, and where BEIJERINCK also physically showed his indefatigability. This was followed by animated talk over a dinner in the company of the two kind sisters, who acted as hostesses. In the evening — at any rate in summer — master and guest went for another walk, deeply immersed, not only in microbiology and botany, but in all the discussion of exact sciences, for BEIJERINCK was profoundly interested in the progress of chemistry, physics and astronomy. He would speak with ardour of BOHR's researches on the model of the atom, and would hold force on the beauties of the works of the astronomers EDDINGTON and JEANS.

Although scientific subjects greatly preponderated in the conversations, it should not be thought that BEIJERINCK never showed any interest in other topics. In the rare moments that he did no longer feel MINERVA's severe look turned towards him, BEIJERINCK became a good-natured and even kind-hearted man.

With regard to literature, however, BEIJERINCK was almost indifferent. BEIJERINCK never ceased to take pleasure in the poems of BYRON, but his pleasure was tinged with a curious sorrow after he learnt that the poet had been unfaithful to his wife. For music BEIJERINCK had no appreciation whatever; its execution he deemed to be extremely fatiguing, and he thought it bad for the scientific achievements of the executants. Characteristic of this attitude was his condemnation of a colleague scientist: "Mr. so and so gets old, he visits concerts." He also maintained his aversion from history: it was the cause of feuds between nations, and the teaching of history in schools should be prohibited! Theology was not in his good graces either. BEIJERINCK could not reconcile the existence of human suffering and misery in the world with the existence of a purposeful Power above Nature. Nature was to him the alpha and omega, and he had so profound a veneration for it that it almost took the place of religion. According to him, life was one with the atom, and ceased with the death of the individual.

The next day was again devoted to lengthy discussions, either walking in the wood or in his grounds, or in the study, and by the time when the visitor came to leave, the conversational quiver of the great scholar was empty, and the guest was tired in body, and limp in mind. Yet the visitor was not dissatisfied, for it was always delightful to hear BEIJERINCK talk; however strange his opinions might sometimes be, there was always something in them leading the hearer to an astonishment which ultimately rose to admiration.

Since BEIJERINCK had been so fortunate as to have such a splendid working-place as the microbiological laboratory at Delft at his disposal, together with the support of a well-trained staff, it is not difficult to understand that he constantly delayed putting into effect his original plan to found a laboratory at Gorssel. Apart from some simple bacteriological experiments, he never returned to regular microbiological researches, but devoted himself entirely to his first love, botany. As was remarked before, the problem of phyllotaxis in con-

Beijerinck in his garden at Gorssel,
at the age of 73.

Beijerinck with his sister and their household companion in 1929.

nection with mathematical considerations lay nearest his heart. This does not mean that he had lost his interest in microbiology [1]; the 300 letters written to his biographer in the course of the ten years granted him at Gorssel bear witness to the vitality of his interest, as they deal almost exclusively with bacteriology. Several times he wrote very enthusiastically about the discovery of bacteriophagy which phenomenon he considered a confirmation of his theory on the *contagium vivum fluidum* [2].

An example of such a letter is reproduced — slightly reduced — in Plate X. Both BEIJERINCK's handwriting and the composition of the letter are characteristic.

Typical for the indestructibility of BEIJERINCK's scientific enthusiasm are the words with which he, at the age of 75, wound up a letter to his successor: "Fortunate are those who now start". This remark has since been written on the wall of one of the rooms in his old laboratory.

Soon after the publication of the Collected Papers had been completed, a long stream of honours began to flow in upon their author; not until then did it become clear to the scientific world what a pioneer BEIJERINCK had been, and in many fields of biology. After being made a corresponding member of the Czecho-Slovakian Botanical Society in February 1922, Denmark accorded him the EMIL CHRISTIAN HANSEN medal on March 10th of the same year. He was invited to come and receive the medal at Copenhagen and lecture there on his life-work. It will hardly be necessary to say that BEIJERINCK had no liking for these ceremonies, and on May 29th, 1922, Professor SÖRENSEN, accompanied by his wife, came to Gorssel to hand him the medal and its money-prize. An illuminated address bearing the signatures of such distinguished scientists as CALMETTE, THEOBALD SMITH, C. O. JENSEN, JOH. SCHMIDT and S. P. L. SÖRENSEN accompanied the medal. A facsimile of this testimony is reproduced in Plate XI.

In the course of the following years BEIJERINCK received many additional distinctions. In 1926 he was elected a Foreign Member of the Royal Society, a nomination which he valued highly, also on account of his veneration for VAN LEEUWENHOEK, who had been the first Dutchman to receive this rare distinction. The Danish and Russian Academies of Sciences had already made him a Foreign Member, as has the British Society for Medical Research. He further became a corresponding member of the Society of American Bacteriologists, of the Deutsche Boden-Gesellschaft, while the Société microbiologique à Leningrad, the Wiener Gesellschaft für Mikrobiologie and the Société pour la Zymologie pure et appliquée à Bruxelles all made him an honorary member. He also was Honorary Chairman

[1] In later years he regarded the United States as the land of the future for microbiology.

[2] See for this also his article: PASTEUR en de Ultra-microbiologie. Verzamelde Geschriften **6**, p. 16.

M het handschrift. Als gij hierop niet
antwoordt zend ik u alleen de drukproef, maar
dat zal nog lang duren want ik schiet in het
geheel niet op door allerlei onoplosbare vragen

Toxin of ultramicrobe. Proef
Gij weet dat het agens der hondsdolheid (een ultramicrobe) langs het zenuwstelsel
naar de hersens gaat en zich daarbij vermeerdert.
Nu gaat de toxin van de tetanus ook door de zenuwen
naar de hersens. De proef zou nu zijn: vermeerdert deze
toxin zich daarbij ook? Ik geloof niet dat dit
ooit geprobeerd is. Denk daarover en vraag aan
bij serum instituut of de proef gedaan kan worden.
Infectie zal in de staart bij i, als het gif (toxin)

bij a is staart afknippen bij
b (B. tetani verspreidt zich niet
maar blijft bij i)
(infectie)
Kan men nu met
de hersens een
nieuwe rat tetanus bezorgen en zoo in serie dan
is de toxin = ultramicrobe. Is het gif dadelijk
uitgewerkt dan niet.
Met vriendelijke groeten
(Veel geluk in 1924)
Dr MW Beijerinck

Facsimile of part of a letter from Beijerinck to one of his collaborators (1924).

of the International Congress of Plant Sciences held at Ithaca (N.Y.) U.S.A. in 1925, while at the same time he was offered an honorary position at the Serum Laboratories of the Veterinary and Agricultural College at Copenhagen, and another at the College for Fermentation Industries at Ghent.

From the beginning of his stay at Gorssel, BEIJERINCK almost completely isolated himself. As has already been said, he never visited Delft again. Soon after his resignation the Amsterdam Academy of Sciences saw him no longer at the meetings. Once he visited the Agricultural College at Wageningen, where his ex-pupil SÖHNGEN was in charge of a new and extremely well-equipped Microbiological Laboratory. Occasionally he went to see his friend HUGO DE VRIES at Lunteren, not far from Gorssel. In vain D'HÉRELLE tried to call on him; but the American soil microbiologist S. A. WAKSMAN was more successful, as we have mentioned. BEIJERINCK, however, was not forgotten by his friends at Delft. In the summer holidays several of them were guests of the BEIJERINCK family, which after September 24th, 1923 consisted only of Prof. BEIJERINCK and his sister Henriëtte, for on that date their sister Johanna died.

On June 14th, 1927 the golden jubilee of his doctorate was commemorated in the auditorium of the Technische Hoogeschool at Delft. After some hesitation, BEIJERINCK decided not to attend this meeting personally, since he was afraid of the fatigues of the journey. On this occasion the Chairman of the Committee, Professor G. VAN ITERSON Jr. gave an address in which he offered to the Technische Hoogeschool a bronze plaquette with BEIJERINCK's portrait, to be fixed in the hall of his old laboratory in the Nieuwe Laan. The plaquette was made by Professor A. W. M. ODÉ. It was formally accepted by the Board of Curators of the Technische Hoogeschool. Hereupon Professor A. J. KLUYVER, BEIJERINCK's successor, made a short address in which BEIJERINCK's great merits were once more outlined. For the speeches made on this occasion the reader is referred to Appendix I [1]).

A small deputation consisting of Dr. F. G. WALLER — BEIJERINCK's old friend since the days of their youth in the Yeast and Spirit Works — Professor VAN ITERSON, and Professor KLUYVER, went to Gorssel. Here the venerable scientist became the recipient of many compliments, and was offered a small reproduction of the plaquette, together with an album containing the names of those who had offered the tribute. The Microbiological Institute of the Agricultural College at Wageningen, received a similar reproduction.

The limelight directed upon BEIJERINCK as a consequence of his golden jubilee also revived interest amongst the general public. This

---

[1]) A few months later also Professor A. J. J. VANDEVELDE held a commemorative address on the occasion of the opening of the course at the College for Fermentation Industries at Ghent.

led amongst others to the publication of an interview with BEIJERINCK by the well-known writer Mrs. W. VAN ITALLIE–VAN EMBDEN, which appeared in the weekly "De Groene Amsterdammer".

This interview — which is reproduced in Appendix J — gives such a vivid impression of BEIJERINCK's personality that it is tempting to make some comment on it. The whole is a typical specimen of BEIJERINCK's conversations as soon as he left the scientific field. Characteristic of BEIJERINCK's statements is the mixture of dissatisfaction, modesty, and self-glorification. For instance, BEIJERINCK emphasizes that neither as student, nor as teacher, nor as professor did he attain what he should have attained according to his own opinion. The remark: "If I had been ambitious, I might have gained some glory," is illustrative of BEIJERINCK's judgement — or better misjudgement — of his own character and achievement. As soon as his interviewer charged him with being too modest, he answered: "Modest? I was a professor born.... I had rediscovered the Mendelian laws, five years before HUGO DE VRIES...." But on the other hand again he criticizes severely his own way of teaching: "Only three years before my retirement did I understand how I had to teach. I had invented the practical course for microbiology. You may call this mere pedantry; I feel it to be the truth."

To all homage BEIJERINCK was almost completely indifferent. He was averse to any ostentation, and one would never have thought that the gloomy solitary man who regularly wandered through the woods of Gorssel with his cape and slouch-hat was such an eminent scholar. In Gorssel he had hardly any acquaintances at all. Yet he founded there a society for scientific lectures, where he spoke on subjects like "Life and Death", "Imagination and Science", but he was as lonely as he had been before. However, when a visitor came, he revived completely, talked incessantly, spoke of old memories and told jokes and anecdotes, so that one might have thought that he was a cheerful man; but hardly had the visitor left, when BEIJERINCK again became reserved and self-contained.

Nobody wished more ardently that he might have a good friend near than his only remaining sister, with whom, owing to her deafness, he could scarcely exchange thoughts.

Plate IX shows BEIJERINCK and his sister, together with their household companion, as they were in this last period of their life.

In April 1929 the first symptoms presented themselves of the disease — intestinal haemorrhage — which was to cause his death. An adenocarcinoma appeared to be present in the rectum. Investigation in the Netherlands Cancer Institute, the van Leeuwenhoekhuis at Amsterdam, showed that the growth was inoperable. BEIJERINCK heard his death sentence with resignation; he was afraid not of death, but of the way in which it would come, and he longed for complete rest. At first the disease gave him great trouble but no pain.

Par la préſente, nous avons l'honneur de notifier que

le Prix
Emile-Chr. Hanſen
pour 1922

a été décerné à

M. le Dr M. W. Beijerinck

ancien Profeſſeur de Microbiologie à l'École ſupérieure Technique de Delft,

en reconnaiſſance des travaux initiateurs accomplis par lui dans le domaine de la Microbiologie,

en particulier

de la fondation du principe de l'application

des Méthodes électives

pour iſolement des microbes,

de ſa culture

du Bacillus radicicola

qui a eu une importance éminente pour le développement et la propagation de la culture des Légumineuſes,

de ſa découverte

de l'Azotobacter chroococcum,

dont les propriétés biologiques particulières ont été largement miſes à profit dans les recherches pratiques ſur le ſol.

Copenhague, le 8 mai 1922.

Theobald Smith

Facsimile of the testimonial accompanying the Emil Christian Hansen Medal, conferred on Beijerinck in 1922.

He hardly dared to eat, spent his days studying, and gradually became thinner, more yellow and weaker. He could not sleep well at night, often fell asleep in his study, and could only walk in his garden. He bore his sorrow quietly and with resignation, and did not complain. He sat lost in thought for hours, looking at the sailing clouds. His thoughts were still always turned to his constant love, science. Problems of the bacteriophage, of the expanding universe, and of phyllotaxis in connection with the constant of EULER, filled his mind. In September 1930 the actual suffering started. He began to dislike salt, did not want to see visitors, and even wrote a note of apology to his friend HUGO DE VRIES who had expressed his desire to visit him. Still greater was the suffering of his only remaining sister, who saw him waste away with great distress and could do so little for her brother, the only relation who was left her.

In November he got very much worse, but his mind remained clear. After December 10th he had to stay in bed. His weakness was great, but the pains were still endurable. On December 22nd he wrote in pencil his last letter to the author, giving his advice how to proceed with the problem of the bacteriophage. The advice ended with the words: "The way is long, but almost certain." The problem of phyllotaxis and all sorts of mathematical questions rushed through his tired brain, and he became very anxious to consult a mathematician on this subject.

Then the great Rest came; and after a day of suffering this high-priest of science died almost imperceptibly on Thursday January 1st, 1931 between 8 and 9 o'clock at night.

Next day the author saw him on his death-bed, hardly changed, his eyes were deep in their sockets. He was like his bereaved sister wrote in her diary: "noble and profound, calm and peaceful, as if thankful that his suffering and struggles had come to an end."

On January 6th the cremation of the mortal remains took place with little ceremony at Westerveld, Velzen, in the presence of his sister, and of numerous friends and admirers. According to BEIJERINCK's wishes no speeches were made [1]). His only nephew, Mr. J. F. BEIJERINCK, offered thanks for the last honours.

The author still hears the words BEIJERINCK once spoke to him, when faced with approaching death: "*Implora aeterna quiete, implora pace.*"

May he have obtained this....

---

[1]) Obituary articles appeared in several newspapers and periodicals. For a list of these see Appendix K.

PART II

BEIJERINCK

THE BOTANIST

BY

G. VAN ITERSON Jr.

CHAPTER IX

STUDIES ON GALLS

BEIJERINCK's first publication was a short paper on the ovipositor of a gall-wasp, *Aphilothrix Radicis* Fabr. [1]). In handing to the writer copies of his first publications BEIJERINCK did not mention his firstling [2]), and it appears that he found it of minor importance. In consulting the publication one admires the exactness and lucidity of the descriptions and the ability with which a beginner in the study of natural sciences exposes his observations and hypothetical suppositions.

The publication is for the greater part of a morphological character, but it contains also remarks on the behaviour of the insect during the act of ovipositing from which it appears that BEIJERINCK already at that time was puzzled by the problems that some years later he solved in such a splendid manner.

Soon afterwards a second publication [3]) appeared in the "Botanische Zeitung" in 1877; it deals with the plant-galls themselves. It is to be considered as a preliminary communication to his doctorate thesis.

In this publication BEIJERINCK set himself the task to draft a system of the *Arthropoda*-galls, choosing especially the development of these galls in the plant organs as basis for their classification. This is not the place to consider whether this classification is still of value; suffice it to state that later works on plant-galls have not made use of it.

It is important, however, for an appreciation of the development of BEIJERINCK's ideas to realize that his studies required him to sift the literature on plant-teratology carefully. The hiatus appearing were supplemented by his own observations, and original remarks are to be found which sometimes go beyond the scope of the publication.

The doctorate thesis which appeared in 1877 covers a larger field. It is entitled: "Bijdrage tot de Morphologie der Plantegallen" (Contribution to the Morphology of Plant-galls) [4]). Here too, the main point lies in the paragraphs dealing with the classification. BEIJERINCK observes that there is no "blood relationship" between plant-

---

[1]) Over de legboor van *Aphilothrix Radicis* Fabr., Tijdschrift voor Entomologie **20,** 186–198, 1876–77 (*Verzamelde Geschriften* **6,** 49–55).

[2]) This may explain why the paper is not to be found in Part **1** of BEIJERINCK's "Collected Papers".

[3]) Botanische Zeitung **35,** 17–22 and 33–38, 1877 (*Verzamelde Geschriften* **1,** 1–7).

[4]) Academisch Proefschrift, Utrecht 1877 (*Verzamelde Geschriften* **1,** 8–80).

galls, and that the only purpose of a classification is to make a survey easier. Furthermore he considered this classification as "the thread which connects his observations".

The doctor thesis contains a large number of interesting observations on numerous gall-formations, illustrated by drawings of the stages of growth and of the anatomical structure. Several of these galls have not been dealt with further in BEIJERINCK's later publications, and the cecidologist may find still some interesting data in this thesis. Of historical importance is the fact that VAN 'T HOFF made a number of determinations at BEIJERINCK's request of the tannin content of *Cynips Kollari*-galls in various stages of development. It appeared that the tannin content of unripe galls, picked at the beginning of August, is very high, and decreases on ripening.

In the years immediately following the publication of his thesis BEIJERINCK's attention was taken up mainly by the study of plant-galls. Field-observations were constantly made in the neighbourhood of Wageningen and further up along the edge of that part of the Netherland province "Gelderland" that is indicated as "de Veluwe". In his house and in his garden too, experiments were started, and these solved the problems which had puzzled him on his botanical excursions. Here is BEIJERINCK's love for experiment awakened!

A deep impression was made on BEIJERINCK by the discovery of the *heterosis* which appeared characteristic of many gall-wasps. B. D. WALSH had already found in 1872 that sometimes a parthenogenetic generation of a gall-wasp is followed by a second generation with male and female insects present, but this publication was unknown to BEIJERINCK at the time of the writing of his thesis. Independently of WALSH, heterosis was rediscovered and published by H. ADLER in Schleswig in 1877, and this publication led BEIJERINCK to publish in 1880 a short communication on the interconnection of *Biorrhiza aptera* and *Teras terminalis* [1]). These were only preliminary studies, and BEIJERINCK himself stated later that the heterogenesis obtained its "wissenschaftliche Begründung" for the first time in 1881 by the "schöne Abhandlung ADLER's: Ueber den Generationswechsel der Eichengallwespen". With tireless exertion BEIJERINCK checked and complemented ADLER's observations during that same year and the following.

In 1882 appeared BEIJERINCK's standard work on galls "Beobachtungen über die ersten Entwicklungsphasen einiger Cynipidengallen". It was published as a communication of the Royal Academy of Sciences in Amsterdam [2]). This paper still commands admiration. With unsurpassable clearness numerous observations on the biology of gall-wasps are described, especially on the method of ovulation, on

---

1) Entomologische Nachrichten und Zoologischer Anzeiger, 1880.

2) Verhandelingen Koninklijke Akademie van Wetenschappen Amsterdam **22**, 1882 (*Verzamelde Geschriften* **1**, 127–281).

the development and the morphological structure of these galls, on the anatomical structure and on the adaptation to external influences. This is all illustrated with not less than 100 original illustrations, most of which are classical examples as to how scientific exactness may be combined with clear arrangement and artistic taste. It is not surprising that several of these drawings have been copied in the most important surveys and textbooks dealing with plant-galls.

We should mention here that in the reproduction of the plates for BEIJERINCK's *Verzamelde Geschriften* these drawings, which appeared originally as lithographs, have suffered severely; cecidologists are advised, therefore, to consult the plates in the original.

These drawings cost BEIJERINCK a great deal of effort; twenty-five years later he still spoke of the fatigue he felt afterwards. It is further of importance that it was BEIJERINCK's wish to add to these uncoloured drawings half-a-dozen coloured ones, for which his sister drew large plates after BEIJERINCK's sketches. These coloured plates are especially attractive [1]). Presumably, the question of cost has prevented the Academy from reproducing them. It was a great disappointment to BEIJERINCK that they were not printed, and he even suggested in 1921 that they be inserted in his *Verzamelde Geschriften*.

Much of what is mentioned in the important treatise is now well known to cecidologists, but the latter are commonly not aware of how much they owe to BEIJERINCK.

It is impossible to give here an adequate outline of the contents. However, we think it well to mention that after a general chapter on the "Cynipiden und ihre Gallen" (BEIJERINCK states that in the five years before the appearance of his publication approximately 50 different *Cynipidae*-galls were investigated in the fresh condition), a restricted number of galls and their inhabitants were subjected to a closer discussion. These elaborate discussions refer to *a.* the Hieracium-gall, *b.* the Terminalis-gall and the Aptera-gall, produced by the same wasp, *c.* the Baccarum-gall and its Folium-gall, *d.* the Megaptera-gall and its Renum-gall, *e.* the Kollari-gall, inhabited by *Cynips Kollari*, an insect of which BEIJERINCK still assumed in 1882 that it reproduced itself exclusively parthenogenetically, and that new Kollari-galls developed under the influence of its eggs, and *f.* the Orthospinae-gall.

Certainly no one who wishes to become thoroughly acquainted with these important galls can ignore BEIJERINCK's work, though his observations require alteration or completion in some points.

We shall specify further only a few of the more important observations made by BEIJERINCK. In the first place it must be recalled that he succeeded in fixing several gall-wasps in the act of ovulation,

[1]) They are kept in the Laboratory for Technical Botany of the University College of Technology at Delft.

by submerging the part of the plant with the ovulating insect in ether, resulting in the immediate death of the insect before withdrawing its ovipositor. By dissecting the plant-organ carefully, the method of oviposition could be determined exactly. Of these observations drawings were made which, especially, are greatly to be admired.

BEIJERINCK even succeeded in dissecting an oak bud on which a specimen of *Biorrhiza aptera* was in the act of oviposition, and in observing, by means of a magnifying glass, the discharge of an egg from the ovipositor. In this way he was able to explain completely the remarkable manner in which the egg passes through the narrow channel of the ovipositor.

From BEIJERINCK's observations still one other point was of especial interest, namely, that the eggs of some gall-wasps are deposited within the plant tissue by means of the ovipositor, but that they are deposited by other gall-wasps on the surface of the undamaged epidermis of the plant organ. The latter happens for instance with the egg which the Folii-wasp, emerging from our common oakleaf-gall, deposits in a small dormant bud at the base of the trunk of the tree. The wasp bores with its short ovipositor through a great number of bud scales, but deposits the egg on the top of the growing point, to which the egg is fastened with a small quantity of mucous secretion.

BEIJERINCK concluded from his observations that the abnormal cell-growth which causes the *Cynipidae*-galls was due neither to an injury, nor to a poison brought into the wound or into the epidermis by the oviposition. The changes of the normal tissue after this opinion start as a result of the "Larvenentwicklung". BEIJERINCK supposed, however, that the stimulation can sometimes become noticeable while the larva is still in the egg. In the first stage of development of the gall no mechanical damage of the tissue by the larva should occur. If the ovum was deposited on the surface of a tissue it should become enclosed by "Umwallung" as a result of the cell division in the neighbouring tissue.

Later on it appeared that BEIJERINCK's notions on these points needed alteration. Presumably, the injury plays a greater rôle than BEIJERINCK supposed, and it is now agreed that the larva produces a larval cavity by sinking into the tissue lying beneath it, which is killed by its secretions, this means that the "Umwallung" is apparent only. This was proved in 1911 by WEIDEL [1]) for the gall of *Neuroterus numismalis*, and in 1914 by MAGNUS [2]) for other *Cynipidae*-galls.

This need of alteration, however, refers only to a part of the development, and later investigators unanimously praise the exact manner in which BEIJERINCK has described the later stages of growth of *Cy-*

[1]) F. WEIDEL, Beiträge zur Entwicklungsgeschichte und vergleichenden Anatomie der Cynipidengallen der Eiche, Flora **102,** 279–334, 1911.

[2]) W. MAGNUS, Die Entstehung der Pflanzengallen verursacht durch Hymenopteren, Jena 1914.

nipidae-galls. MAGNUS, the investigator who after BEIJERINCK studied the origin of *Cynipidae*-galls most fully, refers in his publication, in which he disagrees on several other points with BEIJERINCK's opinions, to the "klassischen Arbeiten BEIJERINCK's". He also gives further evidence of his highest appreciation by beginning the description of his own observations on the Terminalis-gall with the words: "Die Biologie dieser Galle hat durch die bewundernswerten Beobachtungen BEIJERINCKS ihre volle Aufklärung gefunden".

The admiration which even now every expert feels on reading BEIJERINCK's treatise, is due in the first place to the fact that a man was writing who possessed an unusually extensive knowledge of all the subdivisions of the wide field covered by cecidology, *viz.*, ecology, systematics, morphology, teratology, genetics, anatomy, and animal and plant physiology.

BEIJERINCK's next treatise on galls, dating from 1885, dealing with the gall caused by *Cecidomyia Poae* on *Poa nemoralis* [1]), may be considered as a continuation of his great work on galls, and is inspired by the same spirit. This study derives a special importance from the proof that the remarkable appendages developed at the stem of *Poa nemoralis* under the influence of the larva, are real adventitious roots. It is true that they develop at very unusual places on the stem, but they have the structure common to all such roots, and they can develop into normal roots with lateral roots if the gall-bearing part of the stem is planted as a slip. BEIJERINCK attached hereto the far-reaching conclusion: "dass pflanzliche Gewebe, welche die Fähigkeit zur Bildung normaler Organe nicht besitzen, diese Fähigkeit durch die Aufnahme von aussen kommender Stoffe erlangen können".

We mention further a lecture [2]) held by BEIJERINCK in the same year on the subject of galls on *Cruciferae*, in which he gave a survey of these galls only, without going into further detail.

Of much greater importance is BEIJERINCK's treatise of 1888 "Ueber das Cecidium von *Nematus Capreae* auf *Salix amygdalina*" [3]). The importance of this publication is less due to the very careful description of the gall-insect (this time of the family *Tenthredinidae*), of the manner in which the leaf is injured by the insect, and of the structure of the gall, than to the importance of the considerations on the nature of the gall formation.

In this treatise BEIJERINCK dealt with the question as to whether the substance which causes the formation of the gall produces a permanent change in the protoplasm, or whether the change is only

---

[1]) Die Galle von Cecidomyia Poae an Poa nemoralis. Entstehung normaler Wurzeln in Folge der Wirkung eines Gallenthieres, Bot. Zeitung **43,** 306–315 and 320–331, 1885 (*Verzamelde Geschriften* **1,** 386–400).

[2]) Over gallen aan Cruciferen, 1e Bijlage tot de 30e Jaarvergadering der Nederl. Bot. Vereeniging 1885 (*Verzamelde Geschriften* **2,** 1–6).

[3]) Bot. Zeitung **46,** 1–11 and 17–27, 1888 (*Verzamelde Geschriften* **2,** 123–137).

temporary. He concluded that the latter is the case. He points out that excessive nutrition of a plant organ altered by gall formation does not result in enlargement of the cecidium, but that formations are produced of the same type as may occur by excessive nutrition on the unaltered organ. As a typical example, he calls special attention here again to the above-discussed change of the "gall-roots" of the Poae-gall into normal roots; as another example he describes the formation of normal roots within the surviving gall caused by *Nematus viminalis* on *Salix purpurea*.

Parallel to this BEIJERINCK gives examples from which it appears that the properties of the mother plant are still traceable in the gall. "Die sämtlichen Differenzen, durch welche die Blätter von *Rosa canina, R. rubiginosa, R. rugosa* und *R. acicularis* unter sich verschieden sind", were recognized in "den Anhangsgebilden der Bedeguare von *Rhodites Rosae*", when BEIJERINCK produced in his garden the Bedeguar Gall (popularly known as "Robin Pincushion" or "Moss Gall") on the Rose species mentioned with the aid of the gall-wasp. Thus BEIJERINCK is led to the following conclusion: "Es existieren in dem Protoplasma, welches sich auf dem Wege der Cecidiogenese befindet, zwei selbständige Klassen scharf getrennter und grundverschiedener Eigenschaften, nämlich erstens, diejenige der erblichen, dem Cecidium und der Nährpflanze gemeinsamen, und zweitens, diejenige der temporären, nur dem Cecidium eigenthümlichen Charactere. Die letzteren besitzen überhaupt keine Constanz, und vermögen sich keiner einzigen Neubildung, welche von den Geweben des Cecidiums an sich erzeugt werden, aufzuprägen".

In this treatise BEIJERINCK for the first time announces also the hypothesis on the enzymatic nature of the cecidiogenous substances. He found, namely, that Nematus-gall (unlike the *Cynipidae*-galls) continues its development after the egg therein has been killed. He ascribed the development, in this particular case, to poisonous matter passed along with the egg by the mother insect. After making an estimate of the quantity of this poisonous matter, he concluded that an infinitesimal quantity of it must exert an enormous influence on the growth of many cells. It is this circumstance which he expressed by denoting the substance as a "Wuchsenzym". We shall return to this opinion later on (it was contested by MAGNUS in 1903 [1]) and in 1914 [2]).

In the writer's opinion the publication on plant-galls containing the largest number of new ideas is the one published in 1896, again as a communication of the Royal Academy of Sciences at Amsterdam [3]),

---

[1]) W. MAGNUS, Zur Ätiologie der Gallbildungen, Ber. d. deutsch. bot. Ges. **21,** 129–132, 1903.

[2]) W. MAGNUS, Die Entstehung der Pflanzengallen verursacht durch Hymenopteren, Jena 1914.

[3]) Verhandelingen Koninklijke Akademie van Wetenschappen Amsterdam, 2de Sectie, **5,** 1896 (*Verzamelde Geschriften* **3,** 199–232).

entitled "Ueber Gallbildung und Generationswechsel bei *Cynips calicis* und über die Circulansgalle". It is, however, a peculiar fact that the construction of this publication is not as good as that of his previous writings. BEIJERINCK apparently has not been able to completely avoid the inclination to let the numerous difficulties encountered during the solution of this problem exert influence on the report when he came to write up his observations. Two paragraphs on the Circulans-gall were inserted between the other paragraphs which all deal with the Calicis-gall and the Cerri-gall.

In this treatise BEIJERINCK described how he became convinced through circumstantial observations in nature and in his botanical garden, as well as by repeated experiments in the laboratory, that the inhabitants of the "Knopper-galls" which are to be found on the cupule of the acorn of *Quercus pedunculata*, (the gall-wasp of which received the name of *Cynips calicis*) is the agamous generation of an insect which has a second generation which is gamo-genetic. This generation has all the characteristics of another genus of the *Cynipidae*, namely the genus *Andricus*. This second generation supposedly develops in small galls produced by the sting of *Cynips calicis* in the unripe anthers of the Burgundian Oak (*Quercus cerris*). The fecundated females of this *Andricus* species, which BEIJERINCK called *A. cerri*, were supposed to deposit eggs against the inside of the young cupule of *Quercus pedunculata*.

Here the first instance was discovered of a gall-wasp which is heterogenetic as well as heteroecious.

The occurrence of "Knopper-galls" is therefore, according to BEIJERINCK, dependent on the simultaneous presence of both Oak species mentioned at not too great a distance from each other (these gall-wasps are poor fliers). BEIJERINCK calls attention to the fact that this highly valued tanners' material (the "Knopper-galls" are used in the leather factories and for the preparation of tannic acid) is commonly found only on the cupules of *Q. pedunculata* in the countries native to *Q. cerris*, *viz.*, in Austria, Hungary, and south-west Europe, while they are found only sporadically in Germany and the Netherlands. BEIJERINCK's investigation settled, as far as the Netherlands are concerned, that close to the places where this gall was found *Q. cerris* was indeed present, and that one generation of the gall-insect develops thereon. For other countries this does not seem to have been established. Ross mentions for instance on page 71 of "Die Pflanzengallen Bayerns" [1]) that the "Knopper-gall" is found in Bavaria, but that no *Q. cerris* is present there. No one who, just as the writer, has seen BEIJERINCK's convincing experiments, can doubt that further investigation will show that where there are "Knopper-galls" there will also be found specimens of the Burgundian Oak.

---

[1]) H. Ross, Die Pflanzengallen Bayerns und der angrenzenden Gebiete, Jena 1916.

This treatise of 1896 is worth studying not only on account of the important findings which are discussed, but even more so because here BEIJERINCK's general considerations on gall formations reached a culmination. Herein, for instance, full stress is laid on the remarkable fact that the galls show a complete series of "adaptations" which are of use to the insect enclosed (NEGER [1]) spoke of "altruistic adaptations" in similar cases later on), and which adaptations are indispensable since the insect is exposed to attacks from an army of enemies. BEIJERINCK again raises the question as to what mechanism induces the plant-host to make these formations. He once more concluded that there must be some matter which can move freely from one cell to another, and which determines the formation of the developing gall. Since he imagined that the protoplasm does not leave the cell, he supposed that this matter is produced by the larva or is brought along as a poison with the egg by the mother insect. Thus BEIJERINCK comes again to the conception of the co-operation of a growth-enzyme.

BEIJERINCK in this publication draws a further conclusion. He considers it as very probable that there exists no essential difference between the development of meristematic tissues into the full-grown organs of restricted growth and the development of a tissue by cell-division into a gall. When this is right, then with normal ontogenesis too there must be acting a circulating or diffusing substance which determines the form and the physiological function of the developing tissues. The morphological changes caused by this substance which determine the restricted development of the organs should, to a certain extent, act in opposition to the tendency possessed by the cells to transmit their properties unchanged to the daughter cells.

The point of view indicated here is considered of paramount importance by BEIJERINCK not only for the ontogenesis but for phylogenetic development also. The occurrence of mono-cellular variability in this development he believes to be the rule (nowadays this would be called mono-cellular mutation), but he takes the gall formation as proof that multi-cellular variability can also be active.

It is typical of BEIJERINCK that somewhere in the middle of this treatise he deplores the unenthusiastic reception which he feared these novel ideas were to meet. The convincing power — says BEIJERINCK — of the exposition of a law of nature is less determined by the correctness of the law than by the way it harmonizes with current opinions. However, the end of this treatise, which BEIJERINCK, when he wrote it, probably believed to be his last publication on galls, is very cheerful as to the wide prospects which the study of galls opens up. He calls them "formations which cast a new light on the laws of organo-genetics and of variability".

---

[1]) FR. W. NEGER, Biologie der Pflanzen auf experimenteller Grundlage, Stuttgart 1913, p. 533.

In 1902 BEIJERINCK wrote a short communication for the first number of the journal "Marcellia" [1]). For more than 20 years the Kollari-gall had intrigued him. He had stated in 1882, as has been told above, that the inhabitant of *Cynips Kollari*, reproducing parthenogenetically, would produce new Kollari-galls on *Q. pedunculata*, but afterwards he began to doubt his own observations. He repeated the experiments from which he had drawn his conclusions, but the expected results did not emerge. His experience with the Calicis-gall led him, after many unavailing experiments, to isolate a few specimens of *Cynips Kollari* just out of the chrysalis, together with a branch of *Q. cerris*. Within one hour oviposition on the buds was observed: "Alles war einfach und klar; die lange gesuchte Lösung des Rätsels war gefunden". Out of the infected buds of *Q. cerris* there developed small groups of the Circulans-galls inhabited by *Andricus circulans* which galls BEIJERINCK described in 1896. In this case too, therefore, simultaneous existence of heterogenesis and heteroecism is highly probable. One link in the proof is missing here however: BEIJERINCK did not succeed in making the females of *A. circulans* lay eggs in the buds of *Quercus pedunculata*. BEIJERINCK presumed that these powerful insects have the custom to fly about for a long time before copulating. He has not been able to observe the act of copulation. If the presumption is correct, it would explain why the occurence of the Kollari-gall is not bound up with the immediate presence of *Q. cerris*, as appeared to be the case for the formation of the "Knopper-gall".

The last word on this problem has certainly not yet been said.

BEIJERINCK incidentally touched upon the subject of gall formation once more later on, but these later remarks attracted little attention, partly because they appeared in a treatise in which one would not expect to find such a discussion. The passage referred to is of such importance for an appreciation of the development of BEIJERINCK's views that it merits an unabridged reprint. It is found in a treatise published in 1917 entitled "The Enzyme Theory of Heredity" [2]), and reads as follows:

"Long ago already I came to the conviction that the ontogenetic evolution of the higher plants and animals can be best explained by admitting that it is caused by a series of enzymes, for the greater part endo-enzymes, which, becoming active in a fixed succession, determine the morphological and physiological properties gradually manifest in the development. These enzymes in the formation of plant-galls are likewise concerned, and in a study on the galls of the saw-fly *Nematus capreae* on the leaves of *Salix amygdalina*, I gave them the name of "growth enzymes". It is still my opinion that this view is in the main

[1]) Ueber die sexuelle Generation von Cynips Kollari, Marcellia, Padova **1**, 13–20, 1902 (*Verzamelde Geschriften* **4**, 133–138).

[2]) Proceedings of the Section of Sciences, Kon. Akademie v. Wetenschappen Amsterdam **19**, 1275–1289, 1917 (*Verzamelde Geschriften* **5**, 248–258).

correct, but while I formerly thought that the growth enzymes partly derived from the gall-insect, I now recognize that they belong to the plant only, and that the animal does not introduce enzymes into it."

It is apparent that BEIJERINCK's views have matured in 1917, but that the principle underlying his considerations on cecidiogenesis and ontogenesis of organisms has been unchanged. If one considers the importance attached to hormones and auxins in modern morphology, then one realizes that BEIJERINCK's considerations come close to the newer views and that with respect to this problem as well as to many others he was far ahead of his time.

CHAPTER X

MORPHOLOGICAL INVESTIGATIONS ON ADVENTITIOUS FORMATIONS AND REGENERATION PHENOMENA

The younger biologists who are familiar with BEIJERINCK's microbiological work only, and perhaps know also something of his general biological considerations set forth in his later years, will surely be surprised when they study the investigations which occupied him in the years before 1890. Apart from being a specialist in plant-galls, BEIJERINCK in those years appears also to have been a full-fledged plant morphologist.

We have observed before that BEIJERINCK's studies on galls taught him early the value of experimentation in biology. This is probably the explanation of the wide use he made of experiments in his morphological investigations. A great part of BEIJERINCK's botanical work may be regarded as belonging to the field of "experimental morphology".

In Part I of this biography we have seen that circumstances led BEIJERINCK, after the year 1885, to spend his time especially on other problems, and we have observed that plant morphology receded into the background in his studies after 1890. But his interest in it did not disappear completely, and in later years short morphological studies of especial attraction appeared unexpectedly. BEIJERINCK's last paper in fact, belonged to the field of plant morphology. During the last years of his life the problem discussed therein, namely that of phyllotaxis, occupied his mind more exclusively than any of the numerous subjects with which his tireless labours of forty years' duration brought him in contact.

It is strikingly apparent in these morphological studies that BEIJERINCK did not restrict himself to very minute observations and descriptions of structures, or of changes in those structures after experimental interference, but that he drew conclusions from his observations on life-phenonema in general. Repeatedly, ontogenetical and phylogenetical problems were brought forward in these morphological studies, and especially did he trace the fundamental properties of the protoplasm of plant and animal.

Apart from a short paper of the year 1881 "Over het hoefblad" [1]) (On Coltsfoot; *Tussilago Farfara*), which was based especially on a

[1]) Tijdschrift voor Landbouwkunde, Groningen 1881, 5–6, blz. 138–148 (*Verzamelde Geschriften* **1**, 81–89).

publication of P. NIELSON in 1887, but which contains also some very original remarks and observations, we may call an extensive publication in the "Nederlandsch Kruidkundig Archief" of 1882, entitled "Over het ontstaan van knoppen en wortels uit bladen" (On the development of buds and roots from leaves) BEIJERINCK's first purely morphological publication [1]).

This treatise forms the first of a series of studies dealing with the genesis of adventitious organs in the whole vegetable kingdom, in underground organs as well as in those above ground. This explains why the publication is not restricted to the formation of buds and roots from leaves, as the title would suggest. In the introduction, adventitious organs are discussed in general, and even a schematical figure is explained wherein the possible arrangements of such organs on various parts of a plant is represented. It is evident that BEIJERINCK was strongly influenced here by the important study of H. VÖCHTING, "Ueber Organbildung im Pflanzenreich", which appeared in 1878 [2]), but the works of TH. A. KNIGHT, A. BRAUN, A. DE CANDOLLE, CH. DARWIN, J. SACHS and A. DE BARY also appear to have influenced his modes of thought. Below we shall return more specifically to the results of this 1882 study, but we shall first discuss BEIJERINCK's observations in the related field of regeneration.

A treatise "Over regeneratie-verschijnselen aan gespleten vegetatiepunten van stengels en over bekervorming" (On regeneration phenomena of split vegetation-points of stems and on the formation of ascidia), which appeared in 1883, has as its starting point observations which BEIJERINCK made while at the Government Agricultural College at Wageningen on stems of different varieties of *Brassica oleracea acephala* ("choux moellier blanc" of the firm VILMORIN, of Paris) [3]). During the very wet summer of 1882 it was observable that these stems, more than in other years, underwent a process of voluntary splitting along the longitudinal axis, which even included the vegetation-point of the stem. As a result of this, branching of the stem occurred and true regeneration phenomena also showed themselves in leaves which had split when very young. There was also a formation of ascidia.

BEIJERINCK was especially struck with the regeneration symptoms observed in this case, and they led him to experiment on other plants — Cryptogams and Phanerogams — on "the complete or partial return to the original form after removal of part of the tissue". The observations which he made of the recovery, after wounding, of the tops of the youngest leaves at the vegetation point of a *Selaginella*, are very interesting. Although the prosenchymatic reinforcing tissue

[1]) Nederlandsch Kruidkundig Archief, 2e serie, 3e deel, 4e stuk, 1882, p. 438–493 (*Verzamelde Geschriften* **1**, 90–124).
[2]) Bonn 1878.
[3]) Nederlandsch Kruidkundig Archief, 2e serie, 4e deel, 1e stuk, 1883, p. 63–105 (*Verzamelde Geschriften* **1**, 293–317).

and the serrations were not formed again in this case, a certain return to the original form of leaf occurred. This regenerative power of immature leaves was found to be in contrast to the impotence of damaged mature leaves.

Comparative studies of observations on lower and higher animals led BEIJERINCK to the proposing of six rules which should be valid for plants also. These still deserve attention. Of these six we shall cite only two: *a*) the regenerative power is greater, the younger the organism and the tissues, and *b*) the lines along which regeneration occurs coincide in many cases — perhaps in all cases — with the embryonic course of development of the organ.

In connection with what has been mentioned about BEIJERINCK'S ideas on the formation of galls, and what we are going to observe about his ideas on the development of adventitious buds, we wish to emphasize the way in which BEIJERINCK'S ideas in this treatise already coincided with those of SACHS, who supposed that special substances were required to produce special formations. With respect to the formation of ascidia, due to the growing-together of two leaves, or to deformation of a part of one leaf, BEIJERINCK observes, for example: "It appears that one must suppose in all these cases that the direct cause of the anomaly is due to a diminishing of the quantity of the "stem-forming substance" in the vegetation point, which causes at the same time a cessation of the normal relations between this material and the "leaf-forming substance"; in the case of ascidia, which are only appendices of leaves, it must be supposed that a change in the relation between the quantities of the different substances out of which the various parts of the leaf develop acts in a similar way. If the quantity of stem-forming substance is suddenly greatly decreased, then the leaf-forming substance will be present in such a quantity that the whole region around the vegetation point will be occupied by it, resulting in the development of an ascidium".

To forestall the possible criticism that BEIJERINCK found satisfaction in the formulation of hypotheses, we shall quote here his final sentence: "It must be recognized that everything which is stated here about formation of ascidia is of a hypothetical character, and does little to satisfy the mind", of which the last phrase especially is characteristic of a man who is content only when hypothesis is confirmed by experiment.

In the meantime BEIJERINCK'S studies on adventitious organs continued unremittingly; the results were finally set down in an extensive publication appearing as a treatise of the Royal Academy of Sciences in Amsterdam in 1886, under the title "Beobachtungen und Betrachtungen über Wurzelknospen und Nebenwurzeln" [1]). Many of

---

[1]) Verhandelingen der Koninklijke Akademie van Wetenschappen Amsterdam **25**, 1886 (*Verzamelde Geschriften* **2**, 7–121).

his own observations on the development of root-buds and the origin of adventitious roots on different parts of the plant were described therein with great care, and illustrated with 86 especially clear and original drawings. To these observations were added the fairly numerous cases which at that time had already been published in botanical and horticultural literature; the whole was made into an outline which included the entire plant kingdom. Any botanist wishing to get some idea as to how the various plant families show the above-mentioned peculiarities, must still have recourse to BEIJERINCK's treatise, now more than half a century old. Stronger still: any modern biologist desirous of finding the general rule applicable to the many diverse morphological phenomena, and who wishes to completely understand the meaning of it, or who wishes to consider the relation to other manifestations of life, will have his attention held, on reading the introductory discussions and the still more arresting concluding chapter.

The leading motives, which in the first publication of 1882 were stated with a certain reluctance, are emphasized in this more mature treatise. The significance of adventitious organs for the study of ontogenesis is one of them. One needs only to read the statement: "manche Gründe sprechen für die Annahme, dass bei Knospen und Wurzeln die nämlichen Ursachen, welche ihre erste Entstehung veranlassten, auch bei ihrem späteren Austreiben aus einer ruhenden Anlage im Spiele sind". And is not a similar note struck by this thesis: "Die Art und Weise, wie diese Kräfte dabei arbeiten, ist gewiss auf dem Gebiete der Reize zu Hause, und viele Gründe sprechen für die Annahme, dass die ganze Ontogenie auf Nahrungsreizen beruht"?

Another *Leitmotiv* which may be heard repeatedly is the significance of the "transport of matter" for the determination of the place where adventitious growth will occur.

Where the rising sap-stream in the xylem undergoes a change of direction, as a result of encountering specialised structures of the tissue — in undamaged vegetation-points, at the top ends of stems or roots, in axils, at the vertices of the branchings of the leaf-veins, and at the points of origin of the rootlets — there are to be found the places which preferably produce adventitious buds, according to BEIJERINCK.

On the other hand, the points where the plastic nourishment accumulates, or where its movement is retarded or hindered, are preferred for the appearance of adventitious roots. In both cases — as BEIJERINCK points out — one can hardly imagine a more appropriate arrangement, since the young buds, soon to become green and to assimilate independently, must draw upon the water supply on developing; the adventitious roots, however, which may be compared with colourless parasites, must be situated as favourably as possible to receive organic matter produced elsewhere.

But BEIJERINCK is not blind to the fact that still other factors play a part. "Ein unbekannter Einfluss, welcher von den Seitenknospen ausgeht" is certainly one of them. He even came to the conclusion that "zwischen Wurzel- und Knospenbildung eine, gegenseitig förderende Correlation existirt". Especially in the light of modern conceptions on the formation of "auxins", such statements are certainly remarkable.

BEIJERINCK has endeavoured also, by an anatomical study, to indicate the points in the tissues where the adventitious formations first become visible. In every specific case studied by him, he has ascertained whether this formation is effected on callus or "normally", and in the latter case whether they must be called endogenous or exogenous. For the endogenous formations he has completely confirmed the significance which VAN TIEGHEM [1]) and his pupil MOROT [2]) attributed to the *pericycle* (BEIJERINCK, whose treatise was ready before the appearance of MOROT's speaks usually of the *pericambium*, where the term *pericycle* should be preferred).

BEIJERINCK's studies enabled him to draw up rules for the relation between the location of the lateral roots, and thus also of the root buds, and the structure of the vascular bundle in the roots. These rules were corrected in 1888 by VAN TIEGHEM and DOULIOT [3]) in a few minor points only.

One main result of these anatomical observations, namely, that specialized cells are suitable to serve as a starting point for adventitious growth, leads BEIJERINCK back to the consideration of ontogenesis. He formulates the opinion that "jede lebende Zelle die ganze Pflanze neu erzeugen kann", and introduces as a remarkable auxiliary hypothesis that "die Reproductionsmöglichkeit auf der Gegenwart des Zellkernes, die Reproductionsleichtigkeit auf der Beschaffenheit des Cytoplasmas beruhen". He assumes that the nuclei lose something during growth and division, and that this loss halts the divisions, but that whatever is lost may be restored by a vigorous supply of nourishment, among other things. Such a supply would present itself by changes in direction of the transport streams in the plant tissue; thereupon renewed divisions, that is to say, adventitious formations, should occur. One observes here not only how strongly BEIJERINCK was influenced in those days by DARWIN's theory of pangenesis, but also that he applied it in a very original manner.

Of interest are BEIJERINCK's general remarks on observing that many root-buds may be considered to be metamorphosed root-beginnings, while he considers the opposite transformation, *viz.*, buds

[1]) PH. VAN TIEGHEM, Traité de botanique, Paris 1884.

[2]) L. MOROT, Recherches sur le péricycle, Ann. sciences nat. Bot., 6e sér. **20**, 217–309, 1885.

[3]) PH. VAN TIEGHEM et H. DOULIOT, Recherches comparatives sur l'origine des membres endogènes, Ann. sciences nat. Bot., 7e sér. **8**, 1—660, 1888.

into roots, as not seldom occurring. At a time when the homology of organs was the order of the day, such facts drew particular attention.

For various reasons BEIJERINCK's phylogenetical considerations, given at the conclusion of his elaborate treatise, are the most attractive part of his paper. He treats therein the question as to how much light his observations throw upon the methods by which stem and root of higher plants have evolved in the course of time. In the first place he contrasts the two theories on the development of the stem: the stem should be developed from the "Blattbasen", or the stem and the leaf should be considered as homologous to a "thallus".

The former conception was first carefully considered with reference to GOETHE's Metamorphoselehre [1]), to a treatise by DU PETIT-THOUARS [2]) and to that by GAUDICHAUD [3]). It appears that DELPINO's work "Teoria generale della Fillotassi" [4]) which gave a special elaboration of this conception, arrested BEIJERINCK's attention very considerably; undoubtedly the model which BEIJERINCK constructed of the sphere-pile of DELPINO, and which in later years he used to demonstrate repeatedly, dates from this time. BEIJERINCK agrees that the structure of the little stems of mosses and of the young fern-plants point toward the first hypothesis, and especially toward DELPINO's elaboration of it. Yet he rejects this hypothesis, referring among other things to C. DE CANDOLLE's observation of 1881 that the youngest leaves at the vegetation-point show neither an arrangement according to DELPINO's "Blattstandsäule", nor a shifting, as accepted by DELPINO, but that they appear from the first moment with the final phyllotaxis.

The second conception, the thallus theory, is more attractive to BEIJERINCK, and he imagines that higher plants descend from "liverwort-like" ancestors. The often-occurring double-rowed phyllotaxis reminds one of the bilateral thallus of such ancestors. Even in some Orders of which most of the species show spiral-arrangements of the leaves, some "thallous" species occur. BEIJERINCK believes that the transition of the bilateral phyllotaxis into the spiral types which should have occurred in phylogenesis during a later stage of development of the stem, must be viewed in the light of the theory of AIRY. This investigator thought that such higher systems of phyllotaxes are adaptations to the small space available for lateral organs in the buds. AIRY illustrated such a transition by fixing wooden balls to a stretched rubber band, so as to make them conform to a double-rowed arrangement of leaves at a stem, and then letting the band contract, whereupon spiral arrangements actually occurred.

[1]) J. W. VON GOETHE, Versuch über die Metamorphose der Pflanzen, Stuttgart 1831.

[2]) R. DU PETIT-THOUARS, Essai sur la végétation considérée dans le développement des bourgeons, Paris 1809.

[3]) C. GAUDICHAUD, Recherches gén. sur l'organographie, la physiologie et l'organogénie des végétaux, Mém. de l'Acad. des sciences, Paris 1841.

[4]) Genua 1883.

After the development of the stem, BEIJERINCK discussed that of the root. For BEIJERINCK there was no doubt that the root must be considered as a metamorphosed stem: the occurrence of the central cylinder in both suggests this strongly, in his view. In this connection BEIJERINCK considers of importance the occurrence of adventitious buds on the stems as well as on the roots. He sees in his observations on these and other adventitious formations, a confirmation of the conception that "die Wurzeln erst entstanden sind, nachdem die Gefässpflanzen das Thallus-stadium schon verlassen hatten, und dass sie deshalb nichts anderes als metamorphosierte Blattsprosse sein können". His concluding statement is also remarkable: "Die relativ späte phyletische Entstehung der Wurzeln aus den Sprossen erklärt ferner bis zu einem gewissen Grade den in den vorhergehenden Seiten so vielfach nachgewiesenen directen Uebergang der Wurzelanlagen in Knospen, einen Uebergang, welcher offenbar viel Ähnlichkeit mit Atavismus im gewöhnlichen Sinne besitzt, sich davon aber unterscheidet, dadurch, dass nicht die Sprossform des Urahnes, sondern diejenige der Pflanze selbst erscheint."

It appears from this survey, by its nature incomplete, that this treatise also brought more than could be expected from the title.

CHAPTER XI

STUDIES ON PHYLLOTAXIS

Perhaps no subject has fascinated BEIJERINCK more than the problem of phyllotaxis, which was first attacked in the treatise of 1886 on root buds and adventitious roots. The publications of BRAUN [1]), and of L. and A. BRAVAIS [2]), and of SCHWENDENER [3]) on that subject were studied again and again during the years 1890 to 1900, and they led him to make various constructions and calculations. BEIJERINCK was no mathematician, and he was not able to treat the problem along purely mathematical lines. It is remarkable, however, that mathematics had a strong attraction for him. In his library there was a series of mathematical works, which one would never have expected of a biologist at that time. However, he treated geometrical and even algebraical problems usually along empirical lines, and attempted to find solutions by trial and measurement. Naturally this led very often to serious errors, but with such a man as BEIJERINCK even this method sometimes brought remarkable results.

BEIJERINCK's interest in the problem of phyllotaxis was reawakened by the appearance of the wonderfully illustrated work of A. H. CHURCH, containing many new ideas, entitled "On the Relation of Phyllotaxis to Mechanical Laws" [4]). Herein — in contradistinction to most of the earlier literature — stress was laid on the arrangement of the organs at the growing-point, and less significance was attached to the mature state. BEIJERINCK also considered the mode of development of the leaf-primordial pattern to be of the greatest importance for the solution of the problem.

CHURCH's work led BEIJERINCK to put before his assistant VAN ITERSON the case of three circles, whose diameters decrease in a constant ratio, tangent to each other by pairs, with the problem of discovering the conditions that a fourth circle could be constructed in the space between those given, tangent to all three, and at the same time smaller again than the third by the same ratio. BEIJERINCK ex-

[1]) A. BRAUN, Vergleichende Untersuchung über die Ordnung der Schuppen an den Tannenzapfen als Einleitung zur Untersuchung der Blattstellung überhaupt, Berliner Akademie der Wissensch. 16 Juli, 1830.

[2]) L. et A. BRAVAIS, Essai sur la disposition des feuilles curvisériées, Ann. sciences nat. Bot., 2e sér. **7**, 42–110, 1837.

[3]) S. SCHWENDENER, Mechanische Theorie der Blattstellungen, Leipzig 1878.

[4]) London 1904.

pected that on the continuation of the construction, with successive circles decreasing in the same ratio, an arrangement of logarithmic spirals should result. This should perhaps enable one to put CHURCH's constructions on another basis. VAN ITERSON succeeded in proving mathematically that BEIJERINCK's expectation was right, and this question became the starting point for VAN ITERSON's thesis [1]). It was very difficult to make BEIJERINCK agree with this work and its construction, especially to a complete separation of the mathematical and the morphological sides in the presentation, but in later years he stated spontaneously that this separation was correct. After BEIJERINCK had been established for a number of years in Gorssel, he said at one time that of all his reading this thesis was the work he studied most intensely. Evidence that this was really the case is seen in many computations found after his death, and also in a short publication entitled "Verband tusschen de bladstellingen van de hoofdreeks en de natuurlijke logarithmen" (Relation between natural logarithms and phyllotaxis of the Fibonacci series), which appeared in 1927 [2]).

BEIJERINCK's opinion stated therein has never been completely clear to the writer. In the main it is as follows.

If one draws two helices in opposite directions on the surface of a cylinder placed vertically, in such a way that the one helix makes an angle of inclination whose tangent equals $\sqrt{1/2\,(-1+\sqrt{5})}$, while the other helix is perpendicular to the first one, then it may be proved that consecutive points of intersection of the helices on the surface of the cylinder are placed, with respect to each other, at angles of divergence equal to the limiting angle of the Fibonacci-series (137°30′ 28″). It may also be expressed as follows: the surface of the cylinder is divided by these two helices into rectangular areas whose centres are placed at the said angle of divergence to each other. If one considers the cylinder's surface capped by a hemisphere of the same radius, and constructs thereon the helices at the same inclination, then near the top of the sphere these helices approximate to logarithmic spirals drawn on a plane. These spirals will divide the plane into areas of gradually-diminishing size, which will still have the above-mentioned angle of divergence with each other. BEIJERINCK has given to an area delimited by two logarithmic spirals with these angles of inclination the name of "Folium logarithmicum aureum".

BEIJERINCK supposes that in the ideal case with higher plants the meristematic cell-substance at the surface of the growing-point is distributed in areas such as are indicated above for the top of the hemisphere; each area being a "Folium logarithmicum aureum" but

---

[1]) G. VAN ITERSON Jr., Mathematische und mikroskopisch-anatomische Studien über Blattstellungen nebst Betrachtungen über den Schalenbau der Miliolinen, Jena 1907.

[2]) Verslagen Afdeeling Natuurkunde Koninklijke Akademie van Wetenschappen Amsterdam **36,** 585–604, 1927 (*Verzamelde Geschriften* **6,** 28–45).

all of different size. If one imagines further that subsequently, when the stem develops, *one* leaf arises in each area, then it will be clear that successive leaves will be placed with respect to each other at angles of divergence of 137°30′28″. This will be the divergence too, when the stem has grown out into a cylinder.

The reason why the meristematic substance should often be distributed as described above, but in other cases (for instance in decussate phyllotaxis) follows a quite different pattern, has not been made clear to my mind by BEIJERINCK. Neither did he explain how the "contact spirals" are produced which one may draw through the leaves at the growing-point, these spirals being usually of another type and present in other numbers than the contact spirals of BEIJERINCK's construction.

BEIJERINCK does describe original experiments from which it appears that there sometimes occur stresses in a layer of drying colloidal matter which may lead to orthogonal cracks, resulting in a division of the layer into square areas, but the preference of special "angles of inclination" of the borderlines of the areas in the meristematic cell-substance, which forms the basis of his theory, could not be made plausible by these experiments.

The significance which BEIJERINCK attached to this study and the fact that it took the greater part of his time during the last years of his life may justify my having tried to give an elucidation of this work, which in spite of its shortcomings may certainly be called original and remarkable.

CHAPTER XII

MINOR MORPHOLOGICAL RESEARCHES

In 1885 there appears a short but especially attractive communication on the subject of "Gynodioecie bei Daucus Carota L." [1]), wherein BEIJERINCK shows that the occurrence of gynodioecism has been overlooked up to the present in this common wild flower. Two groups of this plant may be distinguished, which may occur side by side in nature. One of the groups possesses snow-white umbels with a centrally-placed small umbel or central flower of dark brown-red colour. The second group is characterized by greenish-red inflorescences which appear during the blossoming time to be already past their bloom, while in reality they are not, since they continue to have a corolla after fertilization, and the leaves of the corolla enlarge in size even thereafter. The flowers of the first group are normally androgynous; those of the second group possess also completely developed ovules and anthers, with apparently normally-developed pollen, but the anthers of the last mentioned flowers always remain closed. The plants of this latter group are therefore "physiologically female".

It is needless to say that BEIJERINCK elucidated his considerations with neat drawings. Also, he did not restrict himself to a simple description of this, in itself, rather interesting case. He added a general consideration on the value of gynodioecism in the vegetable kingdom. It is of note that he could not consider it of any use. He even stated: "Ja, ich möchte die Eigenschaft der Gynodiöcie der Möhre eben als eine schädliche betrachten, allein nicht so schädlich, dass dadurch die Existenz dieser weit verbreiteten und kraftigen Species bedroht wäre." Here again one is given the impression of a very modern opinion on a problem which biologists have thought about for many years, but on which different opinions have often been given.

BEIJERINCK was further greatly interested in the remarkable forms of some Coniferae classified as "Retinisporae". About 1852 C. KOCH reported that he had obtained *Thuya ericoides* (also called *Retinispora ericoides*), a garden plant imported from Japan as a separate species, from a cutting of *Thuya occidentalis.* However, more attention was drawn to such cases by the publications of L. BEISSNER in 1887 and 1889, wherein the latter established that in these cases "youth forms", which deviate from the main forms, maintain themselves by vegetative growth. Besides these main and youth forms, BEISSNER

---

[1]) Nederlandsch Kruidkundig Archief, 2e serie, 4e deel, 3e stuk, 345–354, 1885 (*Verzamelde Geschriften* **1**, 409–414).

made known intermediate forms also, and he showed that one may obtain such youth and intermediate forms by using as slips shootlets which originate closely above the cotyledons of seedlings.

In 1890 BEIJERINCK[1]) was able to add several cases to BEISSNER's interesting observations. Some of these dealt with the development of branches with youth forms on seedlings after damage by frost, by botanical parasites or by root wounds. In these cases, such branches developed so far away from the cotyledons that without the special circumstances mentioned, normal branches should have developed. BEIJERINCK calls attention here to the significance of such observations for the application of HAECKEL's biogenetical "Grundgesetz" on the development of plants.

Other observations dealt with the possibility of having plants retain their youth forms by poor nourishment, for example by cultivating them as potted plants, and BEIJERINCK observed that potted plants are especially suited for the taking of slips from which "Retinisporae" develop. He presumes that the Japanese originally obtained Retinisporae by means of pot cultivations only, *i.e.*, without taking slips.

Of special note further is BEIJERINCK's suggestion that the Sereh-disease of the sugar cane, which drew especial attention in those years since it threatened the cultivation of cane in Java, might be considered as a deviation of the branches of the cane with respect to the main stem, such as conifers show with "youth forms" in their branches. More interesting still are the considerations related to the question of Retinisporae, on the possibility, anticipated by BEIJERINCK, of obtaining dioecious plants from monoecious plants by means of cuttings.

It is obvious that here also the versatility with which BEIJERINCK treated this subject gave a special stamp to this publication.

If one called on BEIJERINCK in Delft in the early summer, when he frequently spent many hours in his garden, one was sure of being shown the specimens of *Cytisus Adami* which he had planted there, and which possessed an unusually large number of branches of *Cytisus laburnum* and of *Cytisus purpureus*. BEIJERINCK had found, indeed, that if he cut off all branches and made an incision into the main stem of *C. Adami*, many dormant buds would develop thereon which developed a large number of "bud variants", especially of *C. laburnum*.

Of his observations on this remarkable tree, which was observed in 1825 by ADAM at Vitry near Paris, and to which BEIJERINCK's attention was called probably by the study of DARWIN's works, BEIJERINCK has made two short communications. One was published in 1900 [2]), the second in 1908 [3]). When the latter publication appeared,

[1]) L. BEISSNER's Untersuchungen bezüglich der Retinisporafrage, Bot. Zeitung **48**, 517–524 and 533–541, 1890 (*Verzamelde Geschriften* **2**, 283–292).

[2]) On the development of Buds and Bud-variations in Cytisus Adami, Proceedings of the Section of Sciences, Koninklijke Akademie van Wetenschappen Amsterdam **3**, 365–371, 1900 (*Verzamelde Geschriften* **4**, 48–52).

[3]) Beobachtungen über die Entstehung von Cytisus purpureus aus Cytisus Adami, Berichte d. deutsch. bot. Ges. **26a**, 137–147, 1908 (*Verzamelde Geschriften* **4**, 305–312).

BEIJERINCK had not heard of the *chimeras*, which H. WINKLER had shortly before obtained from the bittersweet and the tomato, and which were to lead the *Adami*-problem into a completely new trend.

BEIJERINCK considered *Cytisus Adami* as a hybrid between the above-mentioned *Cytisus* species obtained by grafting, of the kind which H. WINKLER later called "Burdo". He therefore called *Cytisus Adami* a graft-bastard. It is self-evident that for this reason the conclusions drawn by BEIJERINCK on the formation of bud variants cannot be maintained in the light of the more recent knowledge on the nature of the "Propfhybride". This does not preclude the fact that a great number of observations and remarks occur in BEIJERINCK's publications which have retained full significance. It is therefore remarkable that they are quoted only occasionally, and that, for example, in an otherwise very complete survey of the problem by N. P. KRENKE, entitled "Wundkompensation, Transplantation und Chimären bei Pflanzen" [1]) they are not mentioned. From KRENKE's survey it appears that the problem is not yet completely solved, notwithstanding the great deal of work done on it since WINKLER's publications of 1907 and 1908. BEIJERINCK's careful observations may certainly contribute still towards the solution.

To support this claim I mention here only one of his observations. BEIJERINCK determined that the leaves of *Cytisus purpureus* show a reaction which he had described for a few other leaves in 1900 (in a treatise on the formation of indigo [2])), and to which he had given the name "necrobiose reaction". If one heats the top of a leaf of *C. purpureus* for a short time above a flame, practically at once a black band appears at some distance from that top. This must be ascribed to the reaction of enzymes developed from the dying protoplasm (the enzymes are killed at the top) on the constituents of the sap. The same experiment with a leaf of *C. laburnum* does not produce this reaction. A leaf of *C. Adami* shows in the necrobiotic region only a brown coloration which moreover occurs not until a few minutes have passed. BEIJERINCK states that it is possible with this reaction to distinguish small leaves of *C. purpureus*, only a few centimeters long, or still smaller, from those of *C. laburnum* and *C. Adami*.

It is very probable that this reaction could be converted into a microscopical one wherewith the nature of the cell-layers of the bastard may be determined, and that a solution will be reached, in this manner of questions which are still waiting to be answered. We call to mind here that LANGE [3]) and KRENKE (*vide* pp. 639 and 640 of his above-cited work), in their study of periclinal chimerae, made use of the difference in the ability of the cells of the two species to take up dyes.

[1]) Berlin 1933.

[2]) On the Formation of Indigo from the Woad (Isatis tinctoria), Proceedings of the Section of Sciences, Kon. Akad. van Wetensch., Amsterdam **2**, 120–129, 1899 (*Verzamelde Geschriften* **3**, 329–336), and: Further researches on the Formation of Indigo from the Woad (Isatis tinctoria), Ibid. **3**, 101–116, 1900 (*Verzamelde Geschriften* **4**, 1–12).

[3]) F. LANGE, Vergleichende Untersuchungen über die Blattentwicklung einiger Solanum-Chimären und ihrer Elterarten, Planta **3**, 181–281, 1927.

CHAPTER XIII

CROSS-BREEDING EXPERIMENTS

In July 1884 BEIJERINCK gave a lecture [1]) at the Netherlands Agricultural Congress which must have drawn a good deal of attention. Using the work of earlier investigators as a basis — in particular examples and experiments derived from DARWIN's "The Variation of Animals and Plants under Domestication" — BEIJERINCK treated the question as to whether varieties breeding true to type, with better properties than the original varieties, may be produced by crossing of species and of varieties of our cultivated plants, and by selection among the descendants of these "mestizos" obtained by self-fertilization or cross-fertilization [2]). He argues further that among the highest-valued varieties of cultivated agricultural crops (at that time) there may be pointed out a great number which originated from *accidental* cross-breeding, and that this number could be enlarged by *artificial* and systematically performed cross-breeding experiments.

He even concludes that the difficulty in improving the cultivated plants is not so much to be found in the production of new varieties as in the determination of their agricultural value. This leads him to recommend that a Society be founded, for the purpose of not only undertaking such cross-breeding experiments, but also of testing the products in practice.

Presumably, BEIJERINCK visualized a working scheme in which he could join, since from those days dates the beginning of his cross-breeding experiments with cereals which were carried out in Wageningen, where he was aided by his colleague and friend Dr. P. PITSCH and his pupil H. DIJT. After BEIJERINCK moved to Delft, these experiments (under his direction) were continued for some time by his pupil. Only concerning the results of the cross-breeding of *Triticum* species, are we fairly well enlightened; on those with barley varieties and barley species there appeared later, in 1888, a very short notice in the publications of the Kon. Akademie van Wetenschappen [3]), from which may be deduced that a continuation of these experiments would probably have given important results. Apparently BEIJE-

---

[1]) Kunnen onze cultuurplanten door kruising verbeterd worden? Verslag van het Landbouwcongres van 22–25 Juli 1884 te Amersfoort gehouden (*Verzamelde Geschriften* **1**, 359–366).

[2]) BEIJERINCK recommends using the term *mestizo* for the just mentioned bastards, and prefers to speak of "hybrids" and "hybridization" where nowadays the term "cross-breeding of species" is used.

[3]) Over kruisingsproeven met kultuurgerst, Versl. en Meded. Kon. Akad. v. Wetensch., Afd. Natuurk. Amsterdam 3de Reeks, **5**, 202, 1888 (*Verzamelde Geschriften* **2**, 189).

RINCK concluded in that year that he had to end these experiments. Either the conditions in Delft were unfavorable for cross-breeding experiments, or BEIJERINCK's attention there was taken up by too many other problems to allow time for such experiments.

On the experiments with *Triticum* species just indicated, there have appeared two small publications in the German language, in 1884 and 1886 respectively; both publications appeared in the Ned. Kruidkundig Archief. The results of only a part of his experiments were given. This may appear from the large collection of wheat ears (unfortunately not in a carefully-preserved condition, and without notes) which BEIJERINCK kept for many years, and which finally came into the possession of the Laboratory for Technical Botany at Delft. The wonderful drawings of flowers and ears of wheat-species present in his collection of plates prove how deep his studies on cereals have been.

After what has been said about BEIJERINCK's lecture, we must call attention to the fact, however, that both publications were written in the first place to throw a light on a scientific problem, namely, the origin of the cultivated species of wheat. By determining which cross-breedings were possible, which succeeded incompletely, and which produced no result, he considered it possible to gain an insight into the relationship of these species. From this BEIJERINCK also expected to gain practical consequences ultimately.

In the first-named publication [1]) BEIJERINCK discusses a bastard obtained by him by cross-breeding *Triticum monococcum* (the "Einkorn") as the mother plant, with *Triticum dicoccum* (the "Emmer") as the father plant. Both plant forms were descended from seed obtained from H. VILMORIN in Paris; of the first species the variety "engrain double", that is, "das doppelte Einkorn", called *Tr. monococcum flavescens* by KÖRNICKE, was used; of the second species, the variety "amidonnier blanc", that is KÖRNICKE's "der weisse, kahle, begrannte Emmer".

The bastards developed into strong plants, rather resembling the mother plant in their vegetative organs, and the male plant in the generative organs. The excellent drawings of the ears, spikelets, and the calyx chaffs, which BEIJERINCK added to the treatise, illustrate many details very clearly. The most important point for BEIJERINCK's considerations was that the bastards appeared to be completely sterile, for he concluded therefrom in his first treatise that *Tr. monococcum* and *Tr. dicoccum* are not related forms. The opinion that these cultivated species were derived from one common wild form — DE CANDOLLE considered this probable for all cultivated cereals — was shaken, therefore, by this observation.

[1]) Ueber den Weizenbastard Triticum monococcum ♀ × Triticum dicoccum ♂, Nederlandsch Kruidkundig Archief, 2e serie, 4e deel, 2e stuk, 189–201, 1884 (*Verzamelde Geschriften* **1**, 401–408).

In the treatise of 1886 [1]) the reciprocal cross-breeding, *viz.*, *Tr. dicoccum*, weisser Emmer ♀ × *Tr. monococcum flavescens*, Körnicke ♂, was discussed in the first place. This cross-breeding succeeded also without difficulty (BEIJERINCK describes exactly the method followed) and the grains of the fertilized mother-plant germinated as well as those obtained with the earlier cross-breeding. The bastard obtained herewith resembled the cross-breeding product described in 1884 very strongly, but small differences in the generative organs were still to be found, to which BEIJERINCK calls special attention (with reference to the work of FOCKE [2])) and which certainly are interesting but cannot be discussed here. The flowers of the bastard developed perfectly normally, and the ovaries also, but there was never found to be any fruit-setting — BEIJERINCK says "zu meiner nicht geringen Verwunderung" — not even on pollination of the bastard with pollen from the mother form, the male form, or with that of *Tr. vulgare, Tr. turgidum*, or *Tr. durum.*

In this treatise of 1886 BEIJERINCK describes furthermore a bastard which he obtained by cross-breeding from *Tr. dicoccum* ♂ with *Tr. monococcum* β *lasiorrachis* Boissier ♀, found wild. He communicates that he received this "wild baeotic wheat" from Mr. H. VILMORIN under the name of *Tr. baeoticum*, but BEIJERINCK doubted the correctness of this indication and changed it into the one just mentioned. The sturdy hybrids obtained were also sterile.

We mentioned above that BEIJERINCK in his first publication repudiated the opinion defended by DE CANDOLLE, among others, that the various species of the cultivated cereals descended from one and the same wild form. In his second treatise he returns, however, to this opinion. Referring to the sterility of the bastard obtained from *Brassica rapa* and *Br. napa*, he considers his observation on the sterility of the wheat bastards obtained as insufficient proof for rejecting the said hypothesis, which attracts him very strongly.

On account of the morphological properties, BEIJERINCK considers the descent of *Tr. monococcum* from the wild *Tr. monococcum* β *lasiorrachis* as practically beyond doubt. With regard to the descent of *Tr. dicoccum*, however, he recognizes that doubt here is justifiable, and he therefore once more discusses at length the various other possibilities in its descent. The significance of a clearer knowledge of this descent he considers especially important, because, to his mind, *Tr. dicoccum* in its turn is to be regarded as the original form of the most important cultivated wheats, namely of *Tr. Spelta, Tr. turgidum, Tr. durum*, and *Tr. vulgare.* He arrives at the conclusion, after these comparisons, that the strongest reasons point toward the above-

[1]) Ueber die Bastarde zwischen Triticum monococcum und Triticum dicoccum, Nederlandsch Kruidkundig Archief, 2e serie, 4e deel, 4e stuk, 455–473, 1886 (*Verzamelde Geschriften* **1**, 415–426).

[2]) W. O. FOCKE, Die Pflanzenmischlinge. Ein Beitrag zur Biologie der Gewächse. Berlin 1881.

mentioned hypothesis of the descent from a common basic form. BEIJERINCK expresses this in the following words at the end of his second treatise: "so muss ich anerkennen, dass die Annahme der Herkunft von *Triticum diccocum* entweder aus einer uralten Culturvarietät von *Triticum monococcum* oder durch die directe Umwandlung irgend einer Form des wilden *Tr. monococcum lasiorrachis*, die Hypothese ist, welche mich auf Grund unserer gegenwärtigen Kenntnisse weitaus am Besten befriedigt."

It is self-evident, that at the present time many of these considerations possess historical value only. If one considers the enormous number of facts which the modern investigator has at his disposal in the study of the descent of our cereals (*vide* E. SCHLIEMANN, Entstehung der Kulturpflanzen [1]), the experimental results and observations that BEIJERINCK could make use of mean very little. It is certainly interesting therefore that he has been right in the main.

First let us state that his observations have been confirmed. Cross-breeding experiments with *Tr. monococcum* have been repeated. About 30 years after BEIJERINCK the significance of these cross-breedings for the solution of the problem has been again recognized; we refer to the synopsis published by BLEIER in 1928 [2]. These experiments, however, often produced negative results, and whenever that was not the case, the bastards were usually completely sterile, as they were in BEIJERINCK's experiments. Only KIHARA [3] communicated in 1924 that he had obtained fruit setting after cross-breeding *Tr. dicoccum* and *Tr. monococcum*.

The origin of the cultivated emmer, *Tr. dicoccum*, has not been completely made clear, notwithstanding the discovery of the wild emmer, *Tr. dicoccoides*, by AARONSOHN [4]. If, however, the strong arguments in favour of the latter species as the original wild form of *Tr. dicoccum* are accepted as conclusive, then one may declare that a common origin of *Tr. monococcum* and of this *Tr. dicoccoides* (and therefore also of *Tr. dicoccum*), from one and the same basic form, is really probable. In the "Schema der Emmer-Ableitung und Verbreitung", present on page 96 of the above cited work of SCHLIEMANN, one finds *Tr. aegilopoides* mentioned as the common ancestral form of the monococcous and dicoccous wheat series.

Finally it should be mentioned that BEIJERINCK's interest in wheat crosses received a new impetus after he became acquainted with

---

1) Dritter Band des Handbuches der Vererbungswissenschaft, herausgegeben von E. BAUR und M. HARTMANN, Berlin 1932.

2) H. BLEIER, Zytologische Untersuchungen an seltenen Getreide- und Rübenbastarden, 5. Intern. Kongr. Vererbungsl., Z. für indukt. Abstamm. u. Vererb. **1**. Suppl. 447–452, 1928.

3) H. KIHARA, Cytologische und genetische Studien bei wichtigen Getreidearten u.s.w. Mem. of the Coll. of Science Kyoto Imper. Univ. Ser. B. **1**, 1–200, 1924.

4) A. AARONSOHN, Über die in Palästina und Syrien wildwachsend aufgefundenen Getreidearten, Verh. K. K. zool. bot. Ges. Wien **59**, 485–509, 1909–1910.

AARONSOHN's above-mentioned find. This is apparent from the fact that he was led to write a short article on the subject in a popular Netherlands journal (De Levende Natuur) [1]).

As it contained a review only of what was known in those days about the origin of the wheat plant we need not to enter in details on its contents.

---

[1]) De ontdekking van den stamvorm der kultuurtarwe, De Levende Natuur, 1 Juni 1911 (*Verzamelde Geschriften* **6,** 80–86).

CHAPTER XIV

INVESTIGATIONS ON GUMMOSIS

As early as the year 1882 BEIJERINCK published a short communication in a little known journal "Sieboldia", with the suggestive title "The gumming disease of fruit trees is contagious" [1]). Therein he stated that he had succeeded (at Wageningen) in producing gummosis in a completely healthy peach tree, by inserting small pieces of gum from a gum-diseased tree under the bark of the healthy specimen. Control experiments with similar wounds, but in which no gum was inserted, showed no gummose formation. In a plum tree also, gummosis could be produced by infecting it with small pieces of gum from diseased peach branches.

BEIJERINCK immediately attached important conclusions to these findings, with reference to the care necessary in horticulture to prevent the spread of gummosis. He also emphasized in this first publication that his observation might become of importance for the obtaining of technically important gums, such as those produced by the *Acacia*'s.

In 1883 there appeared his first detailed publication on the "contagiousness" of the gum disease [2]), and although BEIJERINCK's ideas on this subject later underwent rather important changes, the publication is still more than worth the study. After further investigation and after infection experiments, he came to the result that the transmission of the disease succeeded only when in the pieces of gum there were present spores of a fungus, which his friend Prof. C. A. J. A. OUDEMANS — who, as is well known, devoted himself for many years to the study of fungi — declared to be a new species of the genus *Coryneum*, and to which this mycologist gave the name *C. Beijerinckii* [3]).

Let it be stated here at once that R. ADERHOLD in Berlin (1902) declared this fungus (which he isolated himself, but of which he also received a culture from BEIJERINCK) to be identical with a fungus found often in "Steinobstkulturen" and usually indicated as *Clasterosporium amygdalearum* Sacc., but to which he himself, on grounds of

[1]) De gomziekte der vruchtboomen is besmettelijk, Sieboldia 27 Mei, 1882 (*Verzamelde Geschriften* **1,** 125–126).

[2]) Onderzoekingen over de besmettelijkheid der gomziekte bij planten, Verhandelingen Koninklijke Akademie van Wetenschappen Amsterdam 1883. In BEIJERINCK's *Verzamelde Geschriften* **1,** 321–357 the French translation, which appeared in Archives Néerlandaises des Sciences Exactes et Naturelles **19,** 43–102, 1884 is inserted. Owing to an error, a reference to the earlier papier has been omitted there.

[3]) C. A. J. A. OUDEMANS, Hedwigia, September 5, 1883, Nr. 8.

priority, gave the name *Clasterosporium carpophilum* (Lév.) Aderh. BEIJERINCK in 1906 resigned himself to this change of name, but took up the subject again in 1914, declaring that he preferred to join in with OUDEMANS' authority and to maintain the name *C. Beijerinckii*. It will appear that BEIJERINCK always considered this organism as the most potent cause of the occurrence of gummosis, and anyone who, as the writer, has been permitted to follow BEIJERINCK's experiments, will be convinced that he was right in this matter.

In the treatise of 1883 BEIJERINCK stated the opinion that a primary infection by the said fungus is necessary for the occurrence of the gumming disease in the *Amygdalaceae*. He supposed that *Coryneum* excretes a "ferment" which changes the cell-walls into gum, and that sometimes produces the same change for the cell-walls of the fungus. This enzyme, however, should react further with the protoplasm of living cells in such a way that these cells, sometimes even after they had divided, should produce this same enzyme and should change their cell-walls into gum. In this manner the disease of the infected parts could be transmitted into healthy parts without the latter being reached themselves by the mycelium.

In an extensive final paragraph BEIJERINCK discusses then the reasons that lead him to the conclusion that the formation of gum arabic, also, is caused by an infection with a related fungus. He had received the material necessary for this conclusion while visiting the Kew Botanical Gardens.

We emphasize here that in this treatise there is no question of the isolation of fungi, and thus also no question of infection experiments with pure cultures. According to later communications, BEIJERINCK began with such isolations in 1886, and succeeded in obtaining a highly virulent spore-forming culture of *Coryneum Beijerinckii*. We have already mentioned above that he sent a pure culture to ADERHOLD, who published in 1902 an interesting treatise [1]) on the relation between the gum exudation (Gummifluss) and this organism, in which he completely confirmed BEIJERINCK's conception that the said fungus produces gummosis; ADERHOLD added, however, that further investigation was needed as to whether perhaps also other causes produce gum exudation.

It was presumably this treatise which reawakened BEIJERINCK's interest in the subject of gummosis in the years following 1902, coupled with the fact that a young biologist, A. RANT, a student of Amsterdam University, expressed the desire to study this subject under his direction. In 1906 there appeared a joint publication, and in the same year a dissertation on the subject was offered by RANT in Amsterdam [2]).

[1]) R. ADERHOLD, Über *Clasterosporium carpophilum* (Lév.) Aderh. und dessen Beziehungen zum Gummifluss, Arbeiten der biologischen Abteilung des Gesundheitsamtes **2**, Heft 5, 515, 1902.

[2]) A. RANT, De gummosis der Amygdalaceae, Dissertatie Amsterdam, 1906.

From the title of the first-mentioned article [1]), it appears already that BEIJERINCK's views had broadened during the twenty years in which he had let the problem rest; no doubt the publication of FRANK [2]), and also the treatise of ADERHOLD (*l.c.*), had been of influence.

BEIJERINCK and RANT described carefully how gum-formation occurs through injury of the cambium of the *Amygdalaceae*, and described the changes in the tissue which become visible thereby. They argued, among other things, that the "wound stimulus" makes itself apparent by a gum-formation, covering an area which is limited by a vertically stretched "ellipse", the wound being at the lower focus of this ellipse. Burning, and especially the application of poison (corrosive sublimate) to a wound, increased gummosis greatly. No influence was as strong, however, as an infection with *C. Beijerinckii*, from which BEIJERINCK and RANT concluded that this organism produces a violent poison, with a *traumatic* effect of long duration. The similarity of the results of various causes on this gum-formation then led BEIJERINCK and RANT to the conclusion that — in contradistinction to what BEIJERINCK had thought originally — the cause must not be sought in the *specific* action of the poison produced by the fungus, but that in *all* cases the change in the cells which leads to gum-formation should be the result of the production of toxic substances by the dying cells. Gummosis, therefore, should be a process of "necrobiosis", that is (according to BEIJERINCK's definition), a cell-function which continues after the death of the protoplasm.

The toxic products produced by this protoplasmic death should react with especial intensity with tissue that is still dividing. The walls of the secondary wood which is being formed by the cambium should be especially susceptible of changing into gum. This reaction with the walls should be in itself — according to BEIJERINCK and RANT — nothing other than a normally-progressing process in the tissues, where sometimes only a small quantity of cell-wall material changes into gum and is absorbed, and where in other cases only so much gum is produced that the cells or the vessels are filled therewith. *Gummosis* should therefore mean an excessive activity in the formation of this "cytoclastic" product.

Finally we mention that BEIJERINCK and RANT emphasized the similarity between gum-flow and resin-flow, and that here again they called attention to the practical significance of this process.

Once more — in 1914 — BEIJERINCK returned to the subject of gummosis [3]), and this time also the publication proved to be an en-

[1]) M. W. BEIJERINCK und A. RANT, Wundreiz, Parasitismus und Gummifluss bei den Amygdaleen, Centralblatt für Bakteriologie und Parasitenkunde, II. Abt., **15,** 366–375, 1906. In *Verzamelde Geschriften* **4,** 267–277 the French translation which appeared in Archives Néerlandaises des Sciences Exactes et Naturelles, Sér. 2, **11,** 184–194, 1906 is inserted.

[2]) A. B. FRANK, Die Krankheiten der Pflanzen, 2. Aufl. 1895.

[3]) Gummosis in the fruit of the Almond and the Peachalmond as a process of normal life, Proceedings of the Section of Sciences, Kon. Akademie van Wetenschappen Amsterdam **17,** 810–821, 1914 (*Verzamelde Geschriften* **5,** 168–177).

largement of his field of vision. It is to be regretted that this publication has not received more attention, since one finds therein a summary of his earlier work on the subject, viewed with reference to his later opinions. New and fundamentally important, in the 1914 treatise, is BEIJERINCK's observation that in the over-ripe fruits of the Peach-almond (*Amygdalus amygdalo-persica* Duhamel Dumonceau), and to a lesser degree also in those of the Almond, gum-formation in sieve tubes of the fruit wall occurs as a normal process, whereby the possibility of infections or external wounds of that tissue are excluded. BEIJERINCK supposes that the tender phloem, during ripening and the subsequent drying up, is subjected to stresses which lead to necrobiosis, and therefore to gummosis in this tissue, which generally has little tendency thereto.

This type of wound response should have to be regarded as one of "the normal factors for the development of the fruit", thus being opposed to gummosis as the result of infections, of externally produced wounds, or of poisons introduced. Thanks to the conception of "necrobiosis", BEIJERINCK has been able to combine all these cases under one common heading.

CHAPTER XV

STUDIES ON STARCH, AND PROBLEMS OF COLLOID CHEMISTRY

A short treatise of BEIJERINCK in 1912 on the structure of the starch grain [1]) has contributed much toward making more generally known just what happens in the swelling of a starch grain. This has been very well described in an earlier communication by FRITZSCHE [2]) in 1834, and in the well-known monograph of C. NÄGELI [3]) in 1858 (BEIJERINCK did not know of the observations of these investigators on this subject), and also, in 1908, Mme. Z. GATIN–GRUZEWSKA [4]) had a correct conception of the process, but the simple experiment with which BEIJERINCK elucidated the swelling process is so convincing that the descriptions given by earlier investigators have become of much less importance. This experiment consists of the addition of a solution of tannin to a suspension of swollen starch grains, through which a precipitate is formed inside the starch blisters which shows a Brownian movement. This last fact, especially, removes all doubt as to the liquid nature of the contents of the blisters.

A later short study of BEIJERINCK on "Crystallised Starch" [5]) won less recognition, and the writer is not wholly convinced that what BEIJERINCK considers as "starch crystals" should not in reality be taken as amylodextrin (in WALTER NÄGELI's sense [6])). Yet, the several communications, and especially the accompanying microphotos, are interesting.

It is needless to say that BEIJERINCK's microbiological investigations led him to make himself thoroughly familiar with the properties of the gels, which he used as solid nutrient media for microcultures, and very often also for experiments with enzymes. This explains why BEIJERINCK brought out also a few publications dealing with subjects which one would not expect to have interest for him.

In the "Zeitschrift für physikalische Chemie" [7]) of 1889 there ap-

---

[1]) Structure of the starch-grain, Proceedings of the Section of Sciences, Kon. Akademie van Wetenschappen Amsterdam **14**, 1107–1110, 1912 (*Verzamelde Geschriften* **5**, 21–24).

[2]) J. FRITZSCHE, Über das Amylum, Annalen der Physik u. Chemie **32**, 129–160, 1834.

[3]) C. NÄGELI, Die Stärkekörner, Zürich 1858.

[4]) Z. GATIN-GRUZEWSKA, Sur la composition du grain d'amidon, Comptes Rendus de l'Acad. des sciences **146**, 540–541, 1908.

[5]) Proceedings of the Section of Sciences, Kon. Akademie van Wetenschappen Amsterdam **18**, 305–309, 1915 (*Verzamelde Geschriften* **5**, 195–198).

[6]) W. NÄGELI, Beiträge zur näheren Kenntnis der Stärkegruppe, Leipzig 1874.

[7]) 3. Band, 110–112, 1889 (*Verzamelde Geschriften* **2**, 237–238).

peared a short communication "Ein einfacher Diffusionsversuch", in which BEIJERINCK describes how on the diffusion of a drop of acid placed on a 10 per cent gelatine gel, there appears a depression in the gel at the limit to which the acid has spread. Using this technique, BEIJERINCK was able to make several observations, *viz.*, the diffusion velocity could be studied and measured under the microscope, the liberation of hydrochloric acid due to hydrolysis of ferric chloride, and other observations.

A communication of a colloid chemical nature which appeared in 1896 in the "Centralblatt für Bakteriologie, II. Abt." [1]) was, in fact, of special significance. In this publication BEIJERINCK describes a few experiments with soluble starch obtained by him from potato starch by treatment with hydrochloric acid, and which he used often in his experiments with amylase. It appeared now to him that a solution of this starch in water cannot be mixed with a solution of gelatine to a clear solution, but that the mixing of the two results in an emulsion. By cooling the mixtures of solutions of starch and gelatine in certain proportions, he was able to obtain solid mixed gels, which could be called "künstliche Zellgewebe". The walls of these "spurious tissues" consisted of either starch gel or of solidified gelatine, according to the proportions used.

O. BÜTSCHLI mentions these observations of BEIJERINCK in 1898 on page 251 of his well-known work "Untersuchungen über mikroskopische Strukturen" [2]), with these words: "Dieses für zwei wässerige Lösungen sehr eigentümliche Verhalten, dass mir, offen gestanden, wenig wahrscheinlich vorkam, konnte ich zu meiner Ueberraschung.. ... bestätigen."

In 1910 BEIJERINCK further described the observations just mentioned, and added some similar ones. This time his publication appeared in the "Kolloid-Zeitschrift". We mention here, by the way, that BEIJERINCK had in the meantime observed the same phenomenon, which he described in 1896 for solutions of soluble starch and gelatine also for mixed solutions of gelatine and agar. We further mention that BEIJERINCK defends the conception, in his final considerations, that emulsion-colloids may not be considered simply as droplets of a dispersed phase in a liquid. The final sentence of his publication reads therefore: "Und wenn es sich herausstellen sollte, dass die Eigenschaften der "Emulsionskolloide" nur erklärt werden können, wenn man annimmt, dass die Lösungen derselben aus kleinen wasserhaltigen Substanzmengen bestehen, welche im Dispersionsmittel schweben, dann müssen diese Substanzmengen derart charak-

[1]) Über eine Eigentümlichkeit der löslichen Stärke, Centralblatt für Bakteriologie und Parasitenkunde II. Abt., **2,** 697–699, 1896 (*Verzamelde Geschriften* **3,** 187–188).

[2]) Leipzig 1898.

[3]) Ueber Emulsionsbildung bei der Vermischung wässeriger Lösungen gewisser gelatinierender Kolloide, Kolloid-Zeitschrift **7,** 16–20, 1910 (*Verzamelde Geschriften* **4,** 341–347).

terisiert sein, dass sie sich prinzipiell von den Tröpfchen der mikroskopischen Emulsionen unterscheiden".

It was a long time before BEIJERINCK's observations found the appreciation in colloid chemistry which they merited, but in later years this appreciation was shown. In 1911 TIEBACKX [1]) described a new example of the phenomenon as observed by BEIJERINCK; in 1927 WO. OSTWALD and KÖHLER [2]) devoted a study to another instance, and in 1929 BUNGENBERG DE JONG and KRUYT [3]) added a number of cases, and gave the name of *coacervation* to the phenomenon. Since then it has become of increasing importance in colloid chemistry.

If one examines the present conception as to the nature of "coacervation" (see, for instance, the figure on page 202 of H. R. KRUYT and H. S. VAN KLOOSTER "Colloids" [4])), it will be apparent that BEIJERINCK's conception of a difference between colloidal particles and suspended droplets has been justified.

---

1) F. W. TIEBACKX, Gleichzeitige Ausflockung zweier Kolloide, Kolloid-Zeitschrift **8**, 198–201, 1911.

2) WO. OSTWALD und R. KÖHLER, Über die flüssig-flüssige Entmischung von Gelatine durch Sulfosalizylsäure und über die Beziehungen dieses Systems zur Phasenregel, Kolloid-Zeitschrift **43**, 131–150, 1927.

3) H. G. BUNGENBERG DE JONG and H. R. KRUYT, Coacervation (Partial miscibility in colloid systems), Proceedings of the Section of Sciences, Kon. Akad. v. Wetenschappen Amsterdam **32**, 849–856, 1929.

4) Second Edition, New York 1930.

CHAPTER XVI

PURE CULTURES OF ALGAE

In a lecture held before the "Provinciaal Utrechtsch Genootschap voor Kunsten en Wetenschappen" on June 24th, 1889 [1]), BEIJERINCK reported the successful outcome of his experiments leading to the first pure cultures of green algae ever obtained [2]). In 1890 a larger treatise on the subject appeared under the title "Culturversuche mit Zoochlorellen, Lichenengonidien und anderen niederen Algen" [3]). It will be apparent that in making pure cultures of algae, BEIJERINCK tried out the isolation methods which he had learned in his bacteriological work. He was quickly successful — at least for a number of algae — when he used gelatine media to which no organic nutriments had been added; the cultures were of course exposed to light. He observed, however, that, once isolated, several of the algae grew better when cultivated afterwards on culture media, or in solutions, which did contain organic nutrition; peptone, especially, appeared to act favourably as a nitrogen source. Several of these algae grew excellently even on malt-extract-gelatine without exposure to light.

Once BEIJERINCK was in possession of these pure cultures, he used them for experiments of a nature similar to those in which he had succeeded so well with bacteria. He applied the indigo-white method and also his technique of using luminous bacteria, to demonstrate oxygen formation in red light, and he proved that the algae themselves and yeast-cells, added to the culture, may grow when the suspension is put in red light, even when the solutions do not contain organic substances.

Experiments were then made,. also, to isolate the *Zoochlorellae* of *Hydra viridis,* and those of a green variety of *Stentor polymorphus.* BEIJERINCK had become convinced by the study of the green symbionts of these organisms, that they must be considered identical with one of the green algae which he had isolated (he gave it the name

---

[1]) Over gelatineculturen van ééncellige groenwieren, Aanteekeningen van het verhandelde in de Sectievergaderingen van het Provinciaal Utrechtsch Genootschap K. en W. 35–52, 1889 (*Verzamelde Geschriften* **2,** 227–236).

[2]) H. KUFFERATH in his monograph "La culture des algues" (Paris 1930) mentions that, at the same time as BEIJERINCK MIQUEL succeeded in obtaining pure cultures of algae (diatoms). Without detracting anything from the great merits of the well-known French bacteriologist, it seems that BEIJERINCK has the right of priority, since MIQUEL's paper was published a year later (1890).

[3]) Botanische Zeitung **48,** 725–739, 741–754, 757–768, 781–785, 1890 (*Verzamelde Geschriften* **2,** 293–320).

Chlorella vulgaris). These experiments produced negative results at first, but in a footnote and in a postscript BEIJERINCK communicates that he succeeded in the isolation of the *Hydra*-alga, and that he could identify it as *Chlorella vulgaris*.

The isolation of the gonidia of *Physcia parietina*, which he designated with BORNET as *Cystococcus humicola* Nägeli (later on, with WILLE, as *Chlorococcum humicola*), was easier, and this alga also appeared to thrive only satisfactorily on nutrient-media containing peptone. This led BEIJERINCK to call the Lichens "Doppelparasiten"; the colourless component should profit from the carbon dioxide assimilation of the green symbiont, and the latter from the protein synthesis of the colourless fungus.

In 1893 BEIJERINCK gave a short report on the status of his pure cultures of "niederen Algen" [1]), and in 1898 he communicated that he had finally succeeded in preparing a pure culture of *Pleurococcus vulgaris*, which occurs very widely on the trunks of trees, roofs, and walls [2]). The isolation of *Pl. vulgaris* appeared possible, however, only on an agar plate which has been washed out and freed from all soluble organic matter, and then provided with inorganic salts. Most remarkably, BEIJERINCK was able to ascertain that this organism can adapt itself to organic nutrition.

A publication of 1904 [3]) deals with an alga which BEIJERINCK isolated from "Ulmenfluss", and which he designated as *Chlorella variegata* since the colonies of pure cultures show, next to distinctly green parts, also lighter coloured parts formed by cells which possess less chlorophyll.

Further very interesting illustrated communications on this species of the Family of the *Protococcoideae* are to be found in the classical treatise entitled "Mutation bei Mikroben" [4]), which dates from 1912. It is shown therein that *Chl. variegata* produces two mutants, one of which occurs very regularly on nutrition media containing organic matter. This mutant, designated as *Chlorella variegata aurea*, is characterized by incomplete formation of chlorophyll in the chloroplast. More rarely in cultures, but presumably regularly in nature, a second mutant occurs which BEIJERINCK called *Prototheca Krügeri*, which has completely lost the power to make chlorophyll (not, however, that of forming glycogen in the chloroplast which has become colourless, and which BEIJERINCK designates as "glycophor"). BEIJERINCK feels here that he is justified in assuming a transition from an alga into a fungus, and he states that therewith "die zuerst von SACHS

1) Bericht über meine Kulturen niederer Algen auf Nährgelatine, Centralbatt für Bakteriologie und Parasitenkunde **13,** 368–373, 1893 (*Verzamelde Geschriften* **3,** 21–25).

2) Notiz über Pleurococcus vulgaris, Centralblatt für Bakteriologie und Parasitenkunde II. Abt, **4,** 785–787, 1898 (*Verzamelde Geschriften* **3,** 293–295).

3) Chlorella variegata, ein bunter Mikrobe, Recueil travaux botaniques néerl. **1,** 14–27, 1904 (*Verzamelde Geschriften* **4,** 231–238).

4) Folia Microbiologica **1,** 1–97, 1912 (*Verzamelde Geschriften* **5,** 25–88).

durchgeführte Ansicht des Parallelismus von Algen und Pilzen eine empirische Basis erhalten hat."

In 1902 BEIJERINCK obtained a pure culture also of *Cyanophyceae*, after he had indicated in 1901 how to obtain these organisms from garden soil by enrichment culture in a liquid medium. It became apparent to him, namely, that various *Cyanophyceae* were able to develop in liquids in which only traces of nitrogen were present. If a flask of water from the Delft municipal water supply (this contained approximately 0.42 mg N per liter) to which a small amount of dipotassium phosphate was added (0.02 per cent), was inoculated with garden soil (this contained 0.56 per cent N on the dry matter) and was placed in the light, then therein developed a rich flora which contained many *Cyanophyceae* (*viz.*, species of *Anabaena* and of *Nostoc*). BEIJERINCK considered these organisms as oligonitrophils, and he considered the growth of these cultures so strong that fixation of atmospheric nitrogen had to be assumed. In 1901 he states in a footnote that he will return later to the question as to whether the *Cyanophyceae* themselves fix nitrogen, or whether they do this in symbiosis with other microbes. As appears from his publication of 1902, BEIJERINCK considered the latter the more probable. In 1904 also he states this very distinctly; but attention must be called to the fact that he has not proved this fixation with analytical data.

By spreading the above-mentioned cultures on well-washed plates of agar or silica-gel, to which only 0.02 per cent of dipotassium phosphate had been added, and by cultivating in the light, BEIJERINCK obtained large colonies of bacteria-free *Anabaena*. He adds that his assistant A. VAN DELDEN isolated a blue-green organism on a similar agar medium to which a trace of ammonium nitrate had been added, which organism was related to *Oscillaria*. It is to be regretted that these interesting cultures have not been described more extensively.

Not less important than the isolation of these organisms in pure culture, are the considerations which BEIJERINCK adds to his observations on the possibility that the *Cyanophyceae*, which are apparently satisfied with such simple conditions of life, belong to the oldest organisms on earth. Perhaps even to those which, according to the bold hypothesis of H. E. RICHTER (1865 and 1870), later on independently raised by VON HELMHOLTZ and by WILLIAM THOMPSON, might be distributed through the universe by meteorites. But BEIJERINCK withdrew the latter view in his fundamental publication on "Mutation bei Mikroben" in 1912, and he states that it is much more probable that "abiogenesis" has occurred on earth, be it in earlier geological periods, or that it still occurs.

In a short communication of 1904 [1]) BEIJERINCK describes the

[1]) Das Assimilationsprodukt der Kohlensäure in den Chromatophoren der Diatomeen, Recueil travaux botaniques néerlandais **1**, 28–32, 1904 (*Verzamelde Geschriften* **4**, 239–241).

method by which he obtained *Diatomeae* in pure culture. For this purpose, silica-gel plates, to which had been added dipotassium phosphate and ammonium chloride, appeared especially useful; the technique of preparation of such plates is very carefully described. These pure cultures were used by BEIJERINCK to demonstrate the formation of fat as an assimilation product of these algae.

From the above very condensed survey it will be apparent that in the study of algae also, BEIJERINCK has done pioneer work.

CHAPTER XVII

CONSIDERATIONS ON HEREDITY

Since we restrict ourselves in this Part of the biography to the more purely botanical subjects which had BEIJERINCK's interest, we shall not discuss herein his very important and detailed studies on the variability and the mutability of microbes. Yet we may not pass over this subject completely in the survey of his botanical work, since this work also throws light on heredity in general, and the phylogenetical development in the plant kingdom, an aspect which BEIJERINCK himself has emphasized repeatedly.

On the memorable date of September 29th, 1900, HUGO DE VRIES gave a lecture before the Kon. Akademie van Wetenschappen in Amsterdam, which was to become of historical significance. It was entitled "On the origin of new species of plants", and therein were the first reports of his experiments carried out with the descendants of *Oenothera Lamarckiana*, the seed of which he had gathered from the field. In this lecture, for the first time, the main lines of the "mutation theory" were faintly outlined.

As soon afterwards as Saturday October 27th of that year there followed a lecture by BEIJERINCK "On different forms of heredity variation of microbes" [1]) which he began with these words: "The interesting lecture of Professor HUGO DE VRIES gave at the last meeting of the Academy on the origin of new forms in higher plants, induces me to draw attention to some observations regarding the same subject, in microbes". BEIJERINCK remarks then that, with microbes, it is easier to start from one individual in the making of cultures, that in these cultures many generations succeed each other quickly, that in this case, more easily than with higher plants, large numbers of individuals can be surveyed at one time, and that with many microbes the mutability is great, making them especially suitable for the study of heredity.

It is certainly tempting to cite here from this lecture, but the writer feels that he must restrict himself to one single citation. One of BEIJERINCK's paradoxes was the following: the most important communications of a scientific paper are to be found in the footnotes of the treatise. As a matter of fact, BEIJERINCK's point of view with respect to

---

1) Proceedings of the Section of Sciences, Kon. Akad. v. Wetenschappen Amsterdam **3,** 352–365, 1900 (*Verzamelde Geschriften* **4,** 37–47).

the mutation theory is more clearly expressed in the following footnote of his publication than in the text.

"I perfectly agree with Professor DE VRIES, that the origin of species should often be sought in the almost suddenly produced variants, or mutants, as he calls them. This is also the conclusion to which GALTON has come regarding the races, and to which he referred repeatedly since 1892, the last time, as far as I know, in Nature, vol. 58, p. 274, 1898, in these words: "I have frequently insisted that these sports or "aberrances" (if I may coin the word) are notable factors in the evolution of races. Certainly the successive improvements of breeds of domestic animals generally, as in those of horses in particular, usually make fresh starts from decided sports or aberrances and are by no means always developed slowly through the accumulation of minute and favourable variations during a long succession of generations". Along quite distinct ways GALTON, DE VRIES, and myself, have thus arrived at the same conclusion regarding the probable origin of many races and species. But the great difficulty which lies in the explanation of adaptations, has not been removed, neither by GALTON's "aberrants" DE VRIES' "mutants", nor my "variants"."

The "Proceedings of the Academy" report in a few words that this lecture was followed by a discussion between Professor HUGO DE VRIES and the speaker, in which Professor HUBRECHT also took part. Tradition has it that in this discussion the opinions were sharper opposed than might be thought from the report in the "Proceedings".

A hint of the extent of the differences might also be gained from the fact that BEIJERINCK avoided the use of the word "mutation" until 1912. It was in the title of his extensive study on "Mutation bei Mikroben" which appeared in that year [1]), that he joined in the use of the word. In this treatise also, prospects are opened for the general problem of heredity. A few citations from BEIJERINCK's study may illustrate this.

"Fluktuation und Mutation sind dem Grade nach verschieden. Bei der ersten sind die Sprünge kleiner wie bei der zweiten; die Aussenbedingungen sind beim Zustandekommen der Fluktuation, die Innenbedingungen bei der Mutation überwiegend". "Nach der Genentheorie kann angenommen werden, dass sowohl bei der Mutation wie beim Atavismus Progene in aktive Gene, und umgekehrt Gene in Progene verwandelt werden". "Dass wahrhaft neue Gene bei der Mutation jemals gebildet werden, ist nicht erwiesen, weder bei den Mikroben noch bei den Pflanzen und Tieren. Wenn dieses der Fall zu sein scheint .... so ist doch viel wahrscheinlicher, dass die Progene.... schon in der Stammform gegenwärtig war und durch Atavismus erweckt wurde".

Finally we mention the remarkable publication which is entitled

---

[1]) Folia Microbiologica **1,** 1–97, 1912 (*Verzamelde Geschriften* **5,** 25–88).

"De enzym-theorie der erfelijkheid" (The Enzyme Theory of Heredity) [1]). The writer believes that the cause of the scantiness of the regard evoked by this paper is to be found in the terminology used in it.

BEIJERINCK postulates the following in his treatise. The protoplasm is built up by a large number of factors, which determine the hereditary characteristics of the organism, and which multiply with the cell-division. They received various names and are called — as stated by BEIJERINCK — "differirende Zellelemente (MENDEL), gemmules (DARWIN), biophores, pangenes, genes, character units, heredity units, Mendelian factors, or factors". We emphasize that the nucleus is not taken in consideration herein. BEIJERINCK considers the relation between the protoplasm and the cell nucleus as a separate problem which, however, must be treated *parallel to* the idea just formulated.

There certainly are strong arguments in favour of BEIJERINCK's conception of the "factors". It is in accordance with the older conceptions. In DE VRIES' "Intracellulaire Pangenesis" [2]) one finds in italics, as the main thought: "Das ganze lebendige Protoplasma besteht aus Pangenen; nur diese bilden darin die lebenden Elemente". DE VRIES means with this *protoplasm* the nucleus as well as the cytoplasm. In the definition which W. JOHANNSEN (Elemente der exakten Erblichkeitslehre [3]), 2. Aufl. 1913, S. 143) gives of *gene,* and in which he emphatically states that he therewith concurs with the conception *pangene,* a still wider significance is given to the word *gene,* and it is stated, in spaced letters: "Das Wort Gen ist also frei von jeder Hypothese". JOHANNSEN wishes to express with the conception "genes" only the occurrence of properties "in separable form", so that they can be encountered in different combinations in the gametes and the zygotes. It is remarkable, however, that in the modern study of heredity, notwithstanding the fact that it is historically incorrect, there is a strong tendency to use the conception "genes" exclusively in connection with the nucleus. On p. 508 of the 5th edition of R. GOLDSCHMIDT's excellent "Einführung in die Vererbungswissenschaft" [4]) it is said, for instance: "Wie arbeiten die Gene im Kern — und nur solche kennen wir bisher — mit dem Plasma in dem gesamten jeweiligen System (Eizelle, Keim) zusammen?"

If this difference in conception with respect to "factors" or "genes" is kept in mind, then BEIJERINCK's considerations become clear immediately. Further considerations about his experience on exo- and endo-enzymes convinced him, namely, that enzymes also must be considered as partly living protoplasm (however living protoplasm must not be considered as a simple mixture of enzymes; some enzymes for instance, may first become active in certain stages of the develop-

[1]) Proceedings of the Section of Sciences, Kon. Akademie van Wetenschappen Amsterdam **19,** 1275–1289, 1917 (*Verzamelde Geschriften* **5,** 248–258).

[2]) Jena 1889.

[3]) Jena 1913.

[4]) Berlin 1928.

ment of the cell). This conviction led him to consider whether the "genes of the science of heredity" — *genes* in the original sense — and *enzymes* could not be regarded as identical. His argument, in short, is that this is really the case, and that by the introduction of this supposition, new light is thrown on the nature and on the action of genes during ontogenesis, and also on the occurrence of fluctuating variability and of mutations.

CHAPTER XVIII

BACTERIAL ROOT NODULES

Although BEIJERINCK's microbiological work is amply discussed in Part III of this biography, we should like to give here a brief discussion of his fundamental work on bacterial root-nodules. It was through this research that BEIJERINCK's fame as a bacteriologist was established. Still the work has a definitely botanical side too.

One finds the bacterial root-nodules mentioned already in BEIJERINCK's doctorate thesis [1]) with these words: "Only in a few cases are the galls better known than their causal parasites. This is the case with the root nodules of the *Papilionaceae*". Herewith the cause of BEIJERINCK's later interest in these formations becomes clear, and it also explains his statement: "Die Papilionaceenknöllchen sind Bacteriencecidien".

We recall that by 1888, when BEIJERINCK's classical investigation "Die Bacterien der Papilionaceen-Knöllchen" [2]), appeared, views on the nature of leguminous root nodules had already been promulgated by older investigators. BEIJERINCK mentions a few of these views in a footnote at the beginning of his paper, *viz.*, the observations of WORONIN in 1866 on the presence of living bacteria in root nodules, those of FRANK on the non-occurrence of nodules during the development of *Leguminosae* in sterile soil, those of MARSHALL WARD, on the occurrence of nodules when crushed nodules were added to nodule-free plants grown in sterile soil. But L. HILTNER in his excellent survey in LAFAR's Handbuch der Technischen Mykologie [3]) very properly emphasizes that in 1887 there was still doubt as to the nature of the nodules and that the doubt was strengthened since J. BRUNCHORST had put forward the view that the little bodies in the nodules were protein particles which resembled bacteria, but which should properly be designated as "Bakteroiden". Rightly, HILTNER adds: "Der Umschwung vollzog sich ein Jahr später, also im Jahre 1888, als BEIJERINCK die Pilznatur dieser angeblichen Scheinbakterien dadurch ausser Zweifel stellte, dass er diese aus den Knöllchen abschied und ausserhalb derselben auf künstlichen Nahrböden weiter züchtete".

---

[1]) Academisch Proefschrift, Utrecht 1877 (*Verzamelde Geschriften* **1,** 8–80).

[2]) Die Bacterien der Papilionaceen-Knöllchen, Botanische Zeitung **46,** 725-735, 741–750, 757–771, 781–790, 797–804, 1888 (*Verzamelde Geschriften* **2,** 155–188).

[3]) Dritter Band, Jena 1904–1906, p. 32–34.

As a matter of fact, the just-mentioned treatise is one of BEIJERINCK's masterpieces, not only because of its clear argumentation and the thoroughness with which the morphological as well as the anatomical characteristics of the nodules and their bacteria are described, but also because of the simplicity of the technique applied in the isolation of the bacteria, and of the originality of the methods of studying the physiology of these bacteria. In this first treatise of BEIJERINCK in the domain of "general microbiology" one finds the basis of the application of "auxanography", and use is made of luminous bacteria as reagents for enzymes.

The bacteriological side of BEIJERINCK's investigation has been surveyed in Part III of this book. Here, however, a few points of botanical interest must be made plain, about which BEIJERINCK has quite often been completely misunderstood.

Already in his 1888 treatise BEIJERINCK stated that he had not succeeded in obtaining nitrogen-fixation with cultures of *Bacillus radicicola* (in the beginning he wrote the species name with a capital r) which he had isolated. His opinion on the significance of these nodules was really completely different from what one usually supposes. He suggested that the bacteria produce protein from matter conveyed *by the plant itself*; the bacteroids were to be considered as the reservoirs for this protein, which, in a later stage, would be used by the plant. As an advantage, for the bacteria, of this symbiosis, he indicates that when the nodules decay there occurs a great increase in the number of bacteria, at the expense of the deceased cell tissue. The latter opinion, however, he withdrew in later years.

It seems doubtful whether BEIJERINCK, when writing his treatise, was already acquainted with the extensive report on the experiments of HELLRIEGEL and his co-worker H. WILFARTH, in which nitrogen-fixation by Leguminous plants under natural conditions was convincingly proved. It is certain, however, that BEIJERINCK in 1892 visited HELLRIEGEL in Bernburg, where the latter was experimenting with pure cultures sent to him by BEIJERINCK.

Most botanists and agriculturists will be interested to know BEIJERINCK's view on HELLRIEGEL's experiments. This view was long known to the writer from oral conversations, but BEIJERINCK appears to have hesitated to make it public. His viewpoint has not been expressed, for instance, in the few very short communications of BEIJERINCK on the nodules on the roots of the *Papilionaceae* in 1890 [1]) and 1894 [2]), (interesting observations on these leguminous nodules are also to be found in his lecture before the "Hollandsche Maatschappij der

---

[1]) Künstliche Infection von Vicia Faba mit Bacillus radicicola, Ernährungsbedingungen dieser Bacterie, Botanische Zeitung **48,** 837–843, 1890 (*Verzamelde Geschriften* **2,** 321–326).

[2]) Über die Natur der Fäden der Papilionaceenknöllchen, Centralblatt für Bakteriologie und Parasitenkunde **15,** 728–732, 1894 (*Verzamelde Geschriften* **3,** 49–53).

Wetenschappen" in Haarlem in 1904 [1])). The said point of view may be found, however, in one of BEIJERINCK's latest writings [2]), which certainly must be counted among the most remarkable. This writing deals exclusively with the "significance" for the plant of the bacteria in the nodules.

BEIJERINCK stated emphatically that he did not doubt that proof has been established by HELLRIEGEL, by SCHLÖSING, and by LAURENT (1892), that the nodule bacteria are indispensable for furnishing the *Leguminosae* with the power to fix atmospheric nitrogen. He does doubt seriously, however, whether in the prolonged tests which were done to prove this nitrogen fixation under sterile conditions, after inoculation with a pure culture, there has not occurred some contamination of the soil with other bacteria, among which there may have been free-living nitrogen fixers.

Furthermore, in 1908 BEIJERINCK observed the highly important fact (which seems indeed to be insufficiently known) that nodules which are isolated from the plant can fix no elementary nitrogen; even large quantities of these nodules appear incapable of fixing traces of nitrogen. This fact has been recently confirmed by one of the writer's pupils, G. J. A. GALESTIN [3]).

Finally BEIJERINCK brought forward many observations from nature, from which it appeared that the presence of only a few nodules on the roots of some *Leguminosae* is sufficient for a satisfactory development of these plants. This number was so small, for instance, in a vigorous specimen of *Robinia pseudo-acacia,* which grew in poor heath soil, that, in BEIJERINCK's words, "nobody would attribute to them any direct significance for such a large tree, had not the fixation of nitrogen in the nodules become an inveterate belief".

Rightly, BEIJERINCK concluded in 1918: "Hence, the at present generally accepted explanation of the peculiar behaviour of the *Papilionaceae* cannot be correct. New researches, especially with *Phaseolus*, are desirable".

Much research in this field was done also in the years after 1918, but up till now, BEIJERINCK's problem still awaits an answer.

---

1) L'influence des microbes sur la fertilité du sol et la croissance des végétaux supérieurs, Archives néerlandaises des sciences exactes et naturelles, sér. 2, **9**, VIII–XXXVI, 1904 (*Verzamelde Geschriften* **4**, 249–265).

2) The Significance of the tubercle bacteria of the Papilionaceae for the host plant, Proc. of the Section of Sciences, Kon. Akad. van Wetenschappen Amsterdam **21**, 183–192, 1918 (*Verzamelde Geschriften* **5**, 264–271).

3) Wordt bij de assimilatie van luchtstikstof door Leguminosen elementaire stikstof door de wortelknolletjes geabsorbeerd?, Chemisch Weekblad **30**, 207–209, 1933.

PART III

BEIJERINCK THE MICROBIOLOGIST

BY

A. J. KLUYVER

INTRODUCTION

Even nowadays, the number of people who claim the title of "microbiologist" is very small. This is easily understood if one traces the origin of the scientists who have materially contributed to our knowledge of the microbe world. As a rule it will then become apparent that they have interested themselves in micro-organisms only because they wanted to apply their microbiological experiences to various other branches of science, such as human or animal pathology, phytopathology, industrial or agricultural bacteriology etc. This implies that they prefer to remain physicians, veterinarians, phytopathologists, technologists or agronomists.

Only botanists and zoologists who rightly do not accept any barrier between "higher" and "lower" living organisms, have now and then made disinterested studies of microbes. But the title of "microbiologist" has too narrow a sense for these scientists who hate the restriction imposed by the name, and they avow the unity of living nature by calling themselves "biologists".

If any one, BEIJERINCK was entitled to the qualification of "biologist". Yet, in the second half of his scientific career, he often gave unmistakable proof that he took a special pride in the title "microbiologist". In doing so, BEIJERINCK undoubtedly wished to emphasize that the study of micro-organisms not only calls for special techniques quite foreign to the science of the higher organisms, but also for a special intellectual and mental outlook which is only gradually gained by a continued occupation with microscopic life. In that sense BEIJERINCK was more than a biologist; he was in addition one of the first truly great "microbiologists" of his age and probably of all time.

In the following pages an attempt will be made to justify this eulogy. In the next chapter the circumstances will be set forth which led BEIJERINCK to microbiology, and to his first investigations in this field. In a following chapter a general outline will be given of his development as a microbiologist, whilst in a final chapter the chief contributions made by BEIJERINCK to the science of microbiology will be dealt with in more detail.

CHAPTER XIX

THE BIRTH OF THE MICROBIOLOGIST

In the beginning of the year 1884 BEIJERINCK was an ambitious young botanist who had already attracted world-wide attention by his fundamental contributions to cecidology. Moreover his position as a professor at the Agricultural College of Wageningen seemed to offer many prospects for a harmonious development of his botanical career. In previous years he had devoted himself chiefly to hybridisation experiments on cereals, and there were signs of a growing realization by the educational authorities of the importance of such investigations.

Nothing then seemed to foreshadow any change in the direction of BEIJERINCK's scientific aspirations. Yet an outside agency was to lead to something which at least at first sight seemed nothing short of a revolution.

At that time in Delft an enterprising industrial concern for the production of yeast and alcohol was developing steadily and quickly. The farsighted managing director of this concern, the "Nederlandsche Gist- en Spiritusfabriek", Mr. J. C. VAN MARKEN, realized that further progress of his enterprise might well depend on a more thorough understanding of the properties of the yeast and of the many microscopic enemies which often interfere in its production on a technical scale. For this reason he was anxious to engage on his staff a young biologist with broad scientific qualifications. One need not be surprised that his attention should have been drawn to BEIJERINCK, who, in the autumn of 1884, was offered a position, which not only was very tempting as far as the financial conditions were concerned, but also included the offer of the erection of a new well-equipped laboratory. BEIJERINCK hesitated a long time, but two circumstances made him decide to accept the post. In the first place, the government showed no willingness to meet his wishes as to the building of a new laboratory in Wageningen. Secondly the personality of VAN MARKEN seemed to assure that the new post would offer a large measure of personal liberty, especially as to the kind of work to be undertaken. We shall see that BEIJERINCK's faith in this direction was not betrayed.

Thus it was decided that BEIJERINCK should become a microbiologist. Nevertheless this did not mean that he already was one! There are only slight indications that he had already given any attention to microbiology, during the years he had passed in Wageningen [1]. Yet bacteriology was in a stage of rapid development at that time, as the plate culture method introduced a few years earlier by KOCH led to many successes rendered possible by the isolation of pure cultures.

It is obvious therefore that BEIJERINCK was seriously in want of an initiation into microbiological technique, and DE BARY's laboratory at Strasbourg was deemed to be the right place for this. DE BARY had won a world fame by his fundamental mycological researches, and at the end of his lifetime had centred his interest on the bacteria. His "Vergleichende Morphologie und Biologie der Pilze, Mycetozoen und Bacterien" had just appeared; this was the first treatise in which bacteria were dealt with from the standpoint of the pure biologist. In his obituary of DE BARY, REESS [2] has given a list of all the more prominent scientists who had worked in DE BARY's laboratory, and it is particularly noteworthy that we find amongst those the names of BEIJERINCK, ARTHUR MEYER and S. WINOGRADSKY, all of whom took a leading part in the development of general microbiology during the next quarter of a century.

Although it has been rumoured that BEIJERINCK's fierce character sometimes clashed with the well-earned authority of the German scientist, there is no doubt that it was in DE BARY's laboratory, that the foundations for BEIJERINCK's development as a microbiologist were laid. A hasty visit to E. CHR. HANSEN's laboratory at Copenhagen may have helped him further in getting acquainted with the newer microbiological methods devised by the Danish investigator for the use in fermentation industries, yet there are several indications that BEIJERINCK was not much impressed by the results of this visit.

Here the curtain drops: we have to leave BEIJERINCK alone in his new laboratory in its industrial surroundings, and we can only guess how his initiation into the secrets of the world of yeasts and of bacteria took place.

---

[1] In the introduction to his paper on the contagious character of gummosis he reviews the bacterial plant diseases known until that time, and mentions his unsuccessful attempts to discover bacteria in plant gums.

[2] M. REESS, Ber. deutsch. bot. Ges. **6**, VIII, 1888.

CHAPTER XX

GROWTH AND MATURATION OF THE MICROBIOLOGIST

In the middle of 1885 BEIJERINCK entered upon his post at the "Nederlandsche Gist- en Spiritusfabriek". If one looks in the "Verzamelde Geschriften" for his publications in the years 1886 and 1887 one may be surprised to find several papers dealing with galls, root formation and the *Gardenia* root-disease, showing clearly that his mind was still occupied with the problems which had had his full interest during his stay at Wageningen. It seems probable, however, that these papers dealt chiefly with observations made in that period.

Meanwhile 1887 brought also the first microbiological paper of BEIJERINCK, a lecture held before the "Eerste Nederlandsch Natuur- en Geneeskundig Congres" at Amsterdam on the relation of free oxygen to the vital phenomena of fermentation organisms. In this paper ample proof is given that in the meantime BEIJERINCK had made a thorough study of the historical development of the principal subjects connected with fermentation phenomena. The main feature of the paper, however, is the opinion that — contrary to the view then prevalent — even for strictly anaerobic organisms small quantities of oxygen are indispensable to maintain vital activities.

Whoever might suppose that this first paper of an introductory character would be followed quickly by more detailed communications on the behaviour of fermentation organisms would be mistaken. The year 1888 saw the appearance of a series of highly important papers of a quite unexpected nature, which culminated in the experimental proof that a very special type of bacteria is responsible for the formation of the root nodules of the *Leguminosae.*

At this place we will not enter into a closer consideration of the far reaching importance of this discovery; it may suffice to state that here we have an outstanding contribution to general botany and agriculture made in an industrial laboratory in surroundings which appear most unsuitable for studies of this type. It is obvious that this result must be considered as the direct outcome of BEIJERINCK's previous gall studies, combined with his newly gained experimental abilities in the bacteriological field.

BEIJERINCK succeeded here, where several predecessors failed. The isolation of *Bacillus radicicola*, as BEIJERINCK named the organism in question, may be considered as a bacteriological master-piece for that time. Yet it was performed by a practically self-taught microbiologist who had had only two years of practical experience in the microbiological field!

Undoubtedly encouraged by the sensation which these papers caused amongst botanists and agriculturists, BEIJERINCK unfolded an astonishing productivity in the years which followed. And again, it is most surprising to see that a man who was charged with control and research work in order to promote technical yeast production, was able to spare the energy and time necessary for the solution of several problems of a purely scientific character. It is true that in this connection the very liberal attitude assumed by the management of the "Nederlandsche Gist- en Spiritusfabriek" cannot be too highly praised. But at the other hand it is quite certain that BEIJERINCK did not escape being involved in the numerous troubles inherent in the production of yeast on a technical scale. So for instance in a memorial book entitled "A pilgrimage into yeastland", published by the yeast factory in 1893, we find interesting data regarding the work done by BEIJERINCK to oppose the alarming rumours that pressed yeast could act as a carrier of cholera germs. Besides much experimental work, BEIJERINCK's campaign included several visits to leading bacteriologists and hygienists in England.

Notwithstanding all that, the scientific achievements of BEIJERINCK in his "industrial period" were manifold, and amongst them were several first-rank contributions. We will mention here only his three fundamental papers on the physiology of luminous bacteria, the first and successful application of microbiological methods in the study of unicellular green algae, zoochlorellae and gonidia of lichens — leading to pure cultures of these organisms — his discovery of the remarkable yeast species *Schizosaccharomyces octosporus*, his studies on the butyl alcohol fermentation, those on the micro-organisms of kefir, and on the enzyme lactase, etc.

Moreover, extensive investigations were made on the nutritional requirements of various micro-organisms and new methods for this study were developed, so for instance the so-called "auxanographic method".

Yet there is no doubt that BEIJERINCK's removal in 1895 to more academic surroundings was ultimately felt by him as a liberation. Here, in the new laboratory built according to his own design, seconded by assistants like VAN DELDEN, VAN ITERSON, JACOBSEN and SÖHNGEN, conditions for a further development of the microbiologist were almost optimal.

The characteristic feature of the first three years after the opening of the "Bacteriological Laboratory of the Polytechnical School" in 1897 is that BEIJERINCK had a strong inclination to return to the subjects which had had his interest in the Wageningen period. This manifested itself in an extensive paper on galls, in the publication on mosaic disease in tobacco — which may be considered to mark the beginning of modern virus research — and, finally, in his studies on the formation of indigo, and on the formation of glucosides in species of *Spirea*.

Meanwhile, however, the investigations started at the yeast factory were continued, as appears from the papers on various yeast species, on the pure culture of green algae and on the relation of anaerobic organisms to free oxygen.

An important contribution to general bacteriology was the more or less systematic study on the acetic acid bacteria, which is based largely on the experimental work performed by BEIJERINCK's collaborator D. P. HOYER, who in his doctorate thesis dealt with the subject in more detail.

However, it was only in the period between 1900 and 1910 that BEIJERINCK's genius as a microbiologist came to full maturity.

Almost imperceptibly, a principle came to the fore which will remain for ever one of the foundation stones of microbiological science, *i.e.*, the principle of the accumulation experiment. Whilst until then, the microbiologist who wished to study some special microbe had to rely on his experience regarding the natural occurrence of micro-organisms, and very often also was dependent on mere chance, BEIJERINCK gave a convincing demonstration that in a great many cases it was possible to find the desired germs in nearly every natural material. It is true that as a rule the number of the particular germs in any chosen material will be almost negligibly small so that direct observation or isolation is quite impossible. However, BEIJERINCK was the first to apply consistently the logical idea that by bringing the material in question into a medium, the chemical composition of which was specially adapted to the nutritional requirements of the organism in question, an accumulation must occur which will make subsequent isolation with the aid of the usual pure culture methods an easy task.

When we raise the question at what time this idea has first entered the mind of BEIJERINCK, we have probably to go back to 1894.

The first place in BEIJERINCK's publications where we were able to trace the use of the word "accumulation" ("Anhäufung") is in his paper on sulphate reduction. The discovery and isolation of *Spirillum desulfuricans* were a direct outcome of the application of the said principle [1]). In several later investigations, too, the accumulation principle was more or less consciously applied, yet it was not until 1901, in which year the paper on the urea bacteria was published, that BEIJERINCK insisted on the great significance of the principle. In a footnote the noteworthy remark was made, that its importance

---

[1]) It must be remarked, however, that the first instance of a conscious application of the accumulation principle is to be found in the fundamental investigations of WINOGRADSKY on the nitrifying organisms (1890). In his paper on *Sp. desulfuricans* BEIJERINCK points out the analogy in procedure in the two cases.

should be judged not only from the scientific, but also from the didactic point of view. It is at this point that BEIJERINCK mentions his intention to publish a review of the many experiments of this type which already at that time were regularly carried out in his laboratory [1]).

From that time on BEIJERINCK seems to have been fully aware of the possibilities held out by the so-called "elective culture", and there is no doubt that we owe to this awareness several of his most sensational discoveries. The fundamental researches on oligonitrophilous microbes, leading amongst other things to the discovery of *Azotobacter chroococcum*, were a direct outcome of the enrichment principle. The same can be said of the study made in collaboration with VAN DELDEN "On a colourless bacterium, whose carbon food comes from the atmosphere", *viz.*, *Bacillus oligocarbophilus*, and also of the studies on the thionic acid bacteria, on the lactic acid bacteria, on *Sarcina ventriculi*, etc.

Moreover, in several important papers by BEIJERINCK's collaborators full extension was given to this principle. We may refer in this connection to the papers of VAN ITERSON on denitrifying bacteria and on the bacteria which bring about the aerobic decomposition of cellulose, to those of JACOBSEN on the bacteria which oxidize hydrogen sulphide, sulphur etc., and to those of SÖHNGEN on methane fermentation and on the bacteria oxidizing hydrogen, methane, kerosene, and other hydrocarbons.

By investigations of this character, BEIJERINCK and his school have made a most thorough exploration of the microbe world. In those years one specialized microbe was hardly discovered before an announcement was made of the discovery of another specialized organism with even more remarkable powers!

It would be wrong to leave the impression that the elective method owes its importance only to the fact that it enables the investigator to isolate at any moment any desired type of microbes. BEIJERINCK always emphasized that the results obtained in the enrichment experiments also throw considerable light on the microbial accumulations occurring under natural conditions. In other words, these experiments constitute an important contribution to the ecology of micro-organisms. That herewith one can also get a clearer insight into the rôle of these organisms in the successive processes which have led to the formation of the earth's crust in its present aspect is intimated in several places in BEIJERINCK's papers. Yet, it seems that even nowadays geology is only beginning to awake to the importance of microbial activities in the genesis of many deposits and ores.

---

[1]) Unfortunately BEIJERINCK has never accomplished this task. In 1907, however, a booklet in the German language appeared under the title "Ökologie, Anhäufungen nach BEIJERINCK" by Dr. FERDINAND STOCKHAUSEN. The author who had worked for some time in BEIJERINCK's laboratory had undoubtedly been tempted to this production by the oral expositions of BEIJERINCK.

CHAPTER XXI

A MORE DETAILED APPRECIATION OF BEIJERINCK'S MAIN CONTRIBUTIONS TO MICROBIOLOGY

Although in the previous chapter a general outline already has been given of the eminent services rendered by BEIJERINCK to the science of microbiology, the picture of this great scientist would remain incomplete if no attempt was made to describe with more detail a number of the more important discoveries made by BEIJERINCK in the microbiological field. Herefor the major problems dealt with by BEIJERINCK will successively be passed in review, and since BEIJERINCK's occupations with one and the same problem are often widely separated in time, the survey as a whole will no longer adhere to chronology.

a. *The isolation and investigation of Bacillus radicicola.*

In one of the laboratory note-books (Div. "Bacteria" No. 4), left behind by BEIJERINCK, one finds under the date of May 25th, 1887, a simple entry which on translation reads: "Bacteroids of *Vicia Faba*; those of *Pisum sativum* almost identical. For *Trifolium pratense* small round vesicles." Simple drawings illustrate these statements. The following entry is that of May 31st in which it is reported that on May 26th a small quantity of a ground-up nodule of *Vicia Faba* was sown on a solid culture medium made by adding gelatine to a decoction of the roots of the same plant.

The particular page of the laboratory note-book has been reproduced in Plate XII.

This was the beginning of an enormous amount of experimental work leading to the isolation of *Bacillus radicicola* and to the experimental proof that this bacterium — or closely related varieties and species — is responsible for the formation of the nodules on the roots of *Leguminosae* in general.

Which factors are responsible for this sudden interest of BEIJERINCK for the problem in question? On the one hand it is easily understood that the mystery of the root nodules was already puzzling BEIJERINCK's mind since a long time. We have only to realize that in a former period he was above all a cecidologist, and that the interpretation of the root nodules as a special type of plant gall was at that time unreservedly accepted. In BEIJERINCK's doctorate thesis, which appeared in 1877, the following passage occurs: "Slechts in

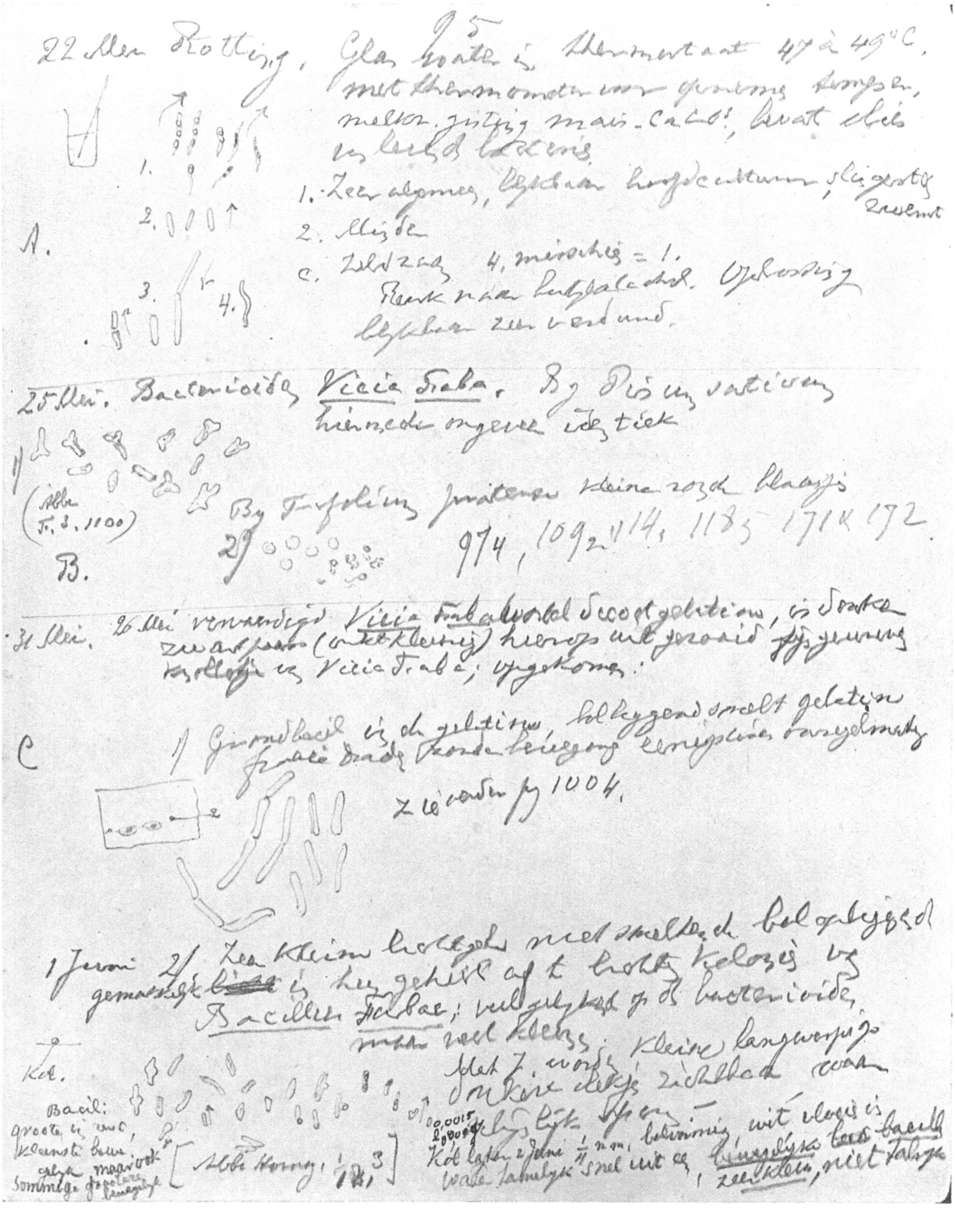

Facsimile of a page of Beijerinck's laboratory note-book (May 22nd–June 1st, 1887), giving his first observations on the root nodule bacteria.

weinig gevallen zijn de gallen nauwkeuriger, de daartoe behoorende parasieten minder goed bekend: dit is het geval met de wortelknolletjes der Papilionaceeën" [1]). Taking into consideration BEIJERINCK's unquenchable thirst for knowledge, it seems probable that during his work at the Agricultural College in Wageningen he would not have lost sight of the problem in question. One might even expect that thus early he would have made various efforts to solve the riddle. Yet no evidence in favour of this view is available [2]), and in any case it appears certain that in his agricultural period BEIJERINCK made no significant advance towards the solution of the question. For we have already seen that at that time BEIJERINCK was not yet a microbiologist, and that certainly he lacked bacteriological experience.

In view of all this, it is most surprising that BEIJERINCK after two years of an industrial career, working in the unfavourable surroundings of the Delft laboratory, suddenly decided to devote a good deal of his time and energy to the subject of root-nodule formation.

Still it is tempting to give some explanation for this unexpected behaviour. The year before, HELLRIEGEL [3]) had published the results of his fundamental investigations which brought convincing proof that the *Leguminosae* possess the exceptional quality of fixing atmospheric nitrogen, but that for that end it is necessary for special bacteria to enter into a symbiotic relationship with the plant, which event then leads to the formation of the root nodules. However, HELLRIEGEL's papers were published in periodicals which were not readily accessible, and BEIJERINCK's attention may well have been drawn to them only by an abstract which appeared in the 1887 volume of the "Centralblatt für Bakteriologie und Parasitenkunde" [4]). The paramount importance of HELLRIEGEL's discovery must certainly have made a great impression on BEIJERINCK's susceptible mind. BEIJERINCK must have felt at once that owing to his newly gained bacteriological experience he was predestined to the task of isolating the as yet unknown causative organism, thus completing the experimental proof of HELLRIEGEL's startling discovery. The scientific passion aroused by this idea made him almost forget that he formed part of an industrial concern, and that it was his task to supervise yeast production. Nor did he evidently pay any attention to the unfavourable conditions under which the work had to be performed.

---

[1]) *Translation:* "Only in a few examples are the galls better known than the parasites; such is, however, the case with the root nodules of the papilionaceous plants".

[2]) Professor ADOLF MAYER, who was intimately connected with BEIJERINCK during the latter's stay in Wageningen, has kindly informed me on my request that he deemed it quite possible that BEIJERINCK already did some experimental work there on the causative organisms of the root nodulus, but that he (A. M.) was unable to find any positive indications in favour of this assumption.

[3]) H. HELLRIEGEL, Tageblatt der 59 Versammlung Deutscher Naturforscher und Ärzte in Berlin, 1886, p. 290; Zeitschr. Ver. Rübenzucker-Industrie deutschen Reichs, **36,** 863, 1886.

[4]) Centralbl. f. Bakt. u. Parasitenk. **1,** 133, 1887.

In the beginning of this paragraph mention has already been made of the fact that his laboratory note-book reveals that he started his investigation on May 25th, 1887. From this date onwards one finds in the note-book a continuous report of observations regarding the bacteroids of various leguminous plants, and also regarding cultural experiments with bacteria obtained out of root nodules. On November 26th, 1887, BEIJERINCK reported the successful outcome of his investigations in the meeting of the "Koninklijke Akademie van Wetenschappen" at Amsterdam. Here for the first time a description of the main properties of the root nodule bacteria was given, and the name of *Bacterium radicicola* proposed [1]).

From his laboratory note-books one sees that BEIJERINCK took up other objects of study soon afterwards. This may explain that almost a year passed before a more detailed publication of the results of BEIJERINCK's investigations on the root nodule bacteria appeared in the "Botanische Zeitung" [2]). It is noteworthy that in this paper no mention is made of HELLRIEGEL's work, although in a footnote to the introduction the most important literature is given. Apparently, BEIJERINCK confined himself here strictly to the bacteriological aspect of the problem, and at that time did not seem it necessary to refer to HELLRIEGEL's primarily agricultural investigations.

It is superfluous to dwell here upon the importance of BEIJERINCK's observations, the paper having become a classic of botanical literature. The circumstantial description of the bacteroids present in the nodules of different *Papilionaceae* has remained unsurpassed. Moreover the paper contains detailed indications for the culturing of the bacterium, the name of which is altered into *Bacillus radicicola* [3]). BEIJERINCK further proves that *Bac. radicicola* is unable to bring about nitrification, and he also reports negative results of experiments intended to demonstrate possible nitrogen fixation by pure cultures of the organism.

Nearly two years later BEIJERINCK returned to this question in a paper which also brings the first direct experimental proof for the nodule forming power of *Bac. radicicola* when brought into contact with aseptically-cultivated *Vicia Faba* seedlings [4]). Here again the nitrogen-fixing power of the pure cultures of the bacterium is denied. However, attention is drawn to the ability of the organism to form a

---

[1]) A detailed abstract of BEIJERINCK's communication was published shortly afterwards in: Versl. en Meded. Kon. Akad. v. Wetensch., Afd. Natuurk., Amsterdam 3de Reeks, **4,** 300, 1888.

[2]) Botanische Zeitung **46,** 725–735, 741–750, 757–771, 781–790, 797–804, 1888. First part published November 16th, 1888.

[3]) BEIJERINCK writes here the specific name: *Radicicola* (with capital R!). That the change in generic name was not due to an altered insight into the systematic position of the organism is clear from the following citation out of BEIJERINCK's 1891 paper on *Bac. radicicola*: "Which bacteriologist will not admit that what we call *Bacillus* nowadays corresponds more or less to the genus "Chaos" of LINNAEUS and comprises essentially different groups?"

[4]) Botanische Zeitung **48,** 837, 1890.

considerable growth at the expense of the very slight amount of nitrogenous substances normally present in water, unpurified sugar, etc.

In the next year a publication appeared in which once more attention is given to the question of a possible nitrogen fixation by the bacterial cultures [1]). It is to be regretted that the title of the paper "Over ophooping van atmospherische stikstof in culturen van *Bacillus radicicola*" — which on translation reads: "On the accumulation of atmospheric nitrogen in cultures of *Bacillus radicicola*" — has led to confusion in so far that it has often been interpreted to imply that at that time BEIJERINCK claimed to have demonstrated the power of the organism to fix free nitrogen. As a matter of fact, BEIJERINCK maintained a very careful attitude towards the results of his experiments, which indeed showed a certain gain of nitrogen in the cultures. BEIJERINCK, however, stressed the possibility that this may have been due to the presence of small amounts of nitrogenous compounds in the air of the laboratory. Experiments undertaken to settle this point were deemed to be inconclusive.

Although the question of nitrogen fixation in pure culture of the root nodule bacteria has since been a matter of much controversy, it may be remarked that BEIJERINCK's critical attitude has afterwards been fully justified by the outcome of various recent investigations on the subject [2]).

The next contribution of BEIJERINCK to the root nodule problem was a short study on the nature of the so-called infection threads often found in the nodules [3]). Experimental proof is given that a close correlation exists between the production of slime in pure cultures of the different strains and the occurrence of the typical infection threads in the corresponding host plants. The conclusion is reached that these threads consist mainly of bacterial mucus, *i.e.*, the slimy cell-walls from which the bacteria themselves have been pressed out more or less completely.

For BEIJERINCK's views regarding the way in which the leguminous plants benefit by the infection with the bacteria the reader is referred to the survey given in Part II of this book (Cf. Chapter XVIII).

b. *Free oxygen in its relation to the vital phenomena of fermentation organisms.*

It is self-evident that BEIJERINCK's work in the yeast factory led

---

[1]) Versl. en Meded. Kon. Akad. v. Wetensch., Afd. Natuurk., Amsterdam 3de Reeks, **8,** 462, 1891. This paper has not been included into the earlier volumes of the "Verzamelde Geschriften"; cf., however, volume **6,** 61.

[2]) Cf.: E. W. HOPKINS, Soil Science **28,** 433, 1929; F. E. ALLISON, Journ. Agric. Research **39,** 893, 1929; M. P. LÖHNIS, Soil Science **29,** 37, 1930.

[3]) Centralbl. f. Bakt. u. Parasitenk. **15,** 728, 1894.

him to make a thorough study of all phenomena connected with fermentation. Amongst these phenomena the way in which free oxygen influences the growth and the fermentative power of the yeast cell has ranked as one of the most important, ever since PASTEUR published his fundamental observations in 1876. If we add that the Delft factory started investigations on the so-called air process of yeast production as early as 1889, and in 1894 introduced this process on a technical scale [1]) it is evident that BEIJERINCK must have had the problem before him during the whole course of his industrial career. At the same time we may presume that BEIJERINCK has made many observations on this point which, on account of their industrial importance, were never published. His publications are largely restricted to the more theoretical aspects of the subject, a circumstance which, at least in a way, enhances the value of these studies.

In a lecture delivered in 1887 during the first meeting of the "Nederlandsch Natuur- en Geneeskundig Congres" BEIJERINCK gave an already authoritive survey of the problem [2]). Herein he made the point that PASTEUR's discovery of the physiological equivalence of fermentation and respiration seemed to have dethroned oxygen as far as its universal indispensability for living organisms is concerned. BEIJERINCK, however, maintained that even for organisms generally considered to be strictly anaerobic, small quantities of oxygen are necessary for the maintenance of life over long periods. This had already been demonstrated in 1880 for ordinary yeast by PASTEUR's pupil COCHIN. BEIJERINCK reported that he had found the same for the strictly anaerobic butyl alcohol bacteria, as also for facultatively anaerobic bacteria, like the lactic acid bacteria and *Bacterium aerogenes*. Therefore, besides its ordinary rôle in respiration, oxygen has an "excitation function", of unknown character, which makes this gas indispensable for all living beings.

In his study on the metabolism of the pellicle forming yeasts, which appeared five years later, BEIJERINCK went so far as to suggest that the significance of gas evolution which so often accompanies anaerobiosis is to be found in the transport of the fermentation organisms to the surface of the medium, thus enabling these organisms to restore their "oxygen reserve" [3]).

In a paper [4]) of 1898 BEIJERINCK returned to the subject. He first of all enounced his opinion that all motile micro-organisms can, on the ground of their behaviour in his "cover glass preparations", be divided into two groups [5]). The cells of the organisms of the first group — to

[1]) F. G. WALLER, Chemisch Weekblad **10,** 635, 1913.

[2]) Handelingen van het Eerste Nederlandsch Natuur- en Geneeskundig Congres, Amsterdam, 1887, p. 34.

[3]) Centralbl. f. Bakt. u. Parasitenk. **11,** 68, 1892.

[4]) Proc. Kon. Akad. v. Wet. Amsterdam **1,** 14, 1898.

[5]) For a detailed description of this "cover glass preparation" method leading to the so-called respiratory figures, the reader is referred to the paper in Centralbl. f. Bakt. u. Parasitenk. **14,** 827, 1893.

which the name of aerophilous organisms was given — seek the highest oxygen tension in the preparation, the organisms of the second, microaerophilous, group evidently prefer lower oxygen tensions. The growth of several so-called obligately anaerobic bacteria was watched both in cultures under the microscope and in shake cultures. In all cases it was observed that optimal proliferation occurred at those spots where low oxygen tensions prevailed. At the end of his paper BEIJERINCK stated explicitly that he did not offer experimental proof for his belief that all living organisms known at that time require free oxygen for their existence. Indeed, the experiments reported demonstrate only that use is made of oxygen in so far as this gas is accessible, and it is admitted that obligately anaerobic bacteria can produce thousands of generations without a renewed contact with free oxygen.

Yet for some facultatively anaerobic bacteria like *B. coli* oxygen — in surprisingly small quantities — is indispensable for the maintenance of life. No explanation could then be offered for this singular fact, and it has not been elucidated in later years.

c. *Studies on luminous bacteria.*

The existence of bacteria capable of emitting light having been demonstrated by PFLÜGER in 1875, some years elapsed before other investigators made a closer study of the various species showing this remarkable property.

In June 1887 FORSTER, who was professor of hygiene at the University of Amsterdam, reported at the meeting of the "Koninklijke Akademie van Wetenschappen" at Amsterdam [1]) the outcome of some investigations on the properties of luminous bacteria, and shortly afterwards his assistant TILANUS also published a paper on the subject [2]).

It seems probable that these publications contributed to the fact that in the next year the industrial microbiologist BEIJERINCK also gave his attention to the group in question. The first entry in his laboratory note-book dealing with luminous bacteria is dated January 12nd, 1888; on that day a sample of luminescent pork received from a Mr. ENKLAAR at Deventer was submitted to a bacteriological analysis. A little later BEIJERINCK seems to have entered into contact with Professor B. FISCHER of Kiel, who had already described several species of luminous bacteria, and the second half of 1888 was mainly devoted to a comparative study of FISCHER's strains and those isolated by BEIJERINCK himself.

As a result BEIJERINCK gave in 1889 a survey of the various species of luminous bacteria then known. He also isolated from water of the North Sea a new species, to which he give the name of *Photobacterium*

[1]) J. FORSTER, Centralbl. f. Bakt. u. Parasitenk. **2,** 337, 1887.
[2]) C. B. TILANUS, Tijdschr. v. Geneesk. **2,** 169, 1887.

luminosum [1]). In the paper BEIJERINCK showed clearly that this organism — which under certain conditions is responsible for the luminescence of the sea water — differs from the ordinary luminous bacteria which practically always can be isolated from sea fish. The statements that the pure cultures sometimes split off non-luminous forms, and that "dissociation" into two different luminous forms may also occur are noteworthy. Regarding the cause underlying the production of light BEIJERINCK remarked that this effect is apparently an incidental consequence of the respiration process: the energy liberated in this process being converted into visible radiation instead of leading to heat production as usual.

In a second communication, which appeared simultaneously with the preceding one, BEIJERINCK dealt extensively with the relations between the luminous bacteria and free oxygen [2]). It was shown that suitable suspensions of luminous bacteria have an even stronger affinity for oxygen than reduced indigo carmine has, since, on adding some reducing agent like sodium hydrosulphite to a suspension containing the indigo dye the light production continued for some time after the dye had been completely converted into its leuco form. In addition arguments were given in favour of the view that oxygen is also an essential excitation agent for the fermentation and the reduction processes caused by *Photobacterium phosphorescens*.

It was not, however, until 1890 that an exhaustive publication of BEIJERINCK's studies on the luminous bacteria appeared [3]). In the first place the various species were divided into two groups, depending on the different nutritional requirements for growth and luminescence. *Ph. phosphorescens* and related species require for their optimal development the presence of a nitrogen-free carbon source, such as sugars and glycerol, besides peptone. On the other hand *Ph. luminosum* and *Ph. indicum* are to some extent inhibited in their development by the addition of such compounds to the peptone media. The discrimination resulted from the application of the elegant auxanographic method described earlier [4]) which can be outlined as follows. A rather large quantity of the cells to be investigated is suspended in an incomplete nutritive medium containing gelatine and by cooling the suspension is solidified in a Petri dish. Then at different spots of the gelatine plate one deposits various chemical substances. If any of these substances supplies the deficient nutritive elements growth will occur in the diffusion field of that substance, and will manifest itself by a local increase in opacity of the plate. The consistent application of this method to various luminous bacteria led BEIJERINCK to a second im-

[1]) Arch. néerl. d. sciences exactes et naturelles **23**, 401, 1889.
[2]) Arch. néerl. d. sciences exactes et naturelles **23**, 416, 1889.
[3]) Versl. en Meded. Kon. Akad. v. Wetensch., Afd. Natuurk., Amsterdam 3de Reeks, **7**, 239, 1890.
[4]) Versl. en Meded. Kon. Akad. v. Wetensch., Afd. Natuurk., Amsterdam 3de Reeks, **6**, 123, 1889.

portant finding. This was that certain compounds (sugars and polyalcohols) had the property of almost instantaneously increasing the luminosity of plates which by "staling" had more or less completely lost the property of phosphorescing. This made him conclude that one had to discriminate between photogenous and "plastic" (*i.e.*, growth promoting) food substances of luminous bacteria. This observation would certainly be interpreted to-day as a strong indication that light production is intimately connected with the respiration processes of the cells, and is independent of proliferation. BEIJERINCK himself arrived at a different conclusion concerning the metabolic process responsible for the light production; this, however, does not detract from the value of these fundamentally important observations.

Mention should be made also of the fact that BEIJERINCK devised several elegant applications of the principles outlined above for the detection of various enzymes. For instance the first experimental proof for the existence of the enzyme lactase (in *Saccharomyces Kefyr*) was adduced from application of these methods. WIJSMAN [1]), working under BEIJERINCK's direction, applied the method successfully in analysing the amyloclastic enzymes present in barley; his findings did not attract much attention at the time, but have since been corroborated by recent investigators [2]).

Ten years later BEIJERINCK published his now well-known observations on the applicability of luminous bacteria for the detection of the traces of oxygen formed in the photochemical reduction of carbon dioxide in green cells [3]). The experiments culminate in the observation that with the aid of this method it is even possible to prove that production of oxygen occurs, when a suspension of chloroplasts, obtained by crushing green leaves and filtering the diluted mass, is illuminated. In a fairly recent survey on photosynthesis it is still remarked that this experiment seems to offer the only example in which it has been possible to prove the occurrence of an — albeit weak — photosynthetical action in the absence of intact living cells [4]). It has, however, to be added that recent investigations seem to prove that this oxygen evolution is not the result of carbon dioxide assimilation, but depends on a photochemical decomposition of some peroxide active in the photosynthetic apparatus [5]).

In the last phase of his career BEIJERINCK returned once more to the subject in question in a paper describing *Photobacterium splendidum*, a still unknown species, responsible for the phosphorescence of the

---

[1]) H. P. WIJSMAN, De diastase beschouwd als mengsel van maltase en dextrinase. Amsterdam, 1889.
[2]) Cf. G. A. van Klinkenberg, Ergebn. der Enzymforschung **3**, 73, 1934.
[3]) Proc. Kon. Akad. v. Wet. Amsterdam **4**, 45, 1901.
[4]) R. EMERSON, Ergebn. d. Enzymforschung **5**, 305, 1936.
[5]) H. KAUTSKY, Die Naturwissenschaften **26**, 14, 1938.

North Sea after hot days in the summer months [1]). BEIJERINCK observed in this species the remarkable phenomenon of aggregation, due to the micro-aerophily of the majority of the individual cells. He also reported the interesting observation — made in collaboration with F. C. GERRETSEN — that luminous bacteria exposed to ultraviolet radiation lose their reproductive function rather quickly, whilst they continue to emit light for several hours [2]). This experiment provided the first and so far the only example of light emission by material derived from luminous bacteria, *i.e.*, in the absence of normal cells capable of reproduction. In a final part, BEIJERINCK discussed many observations regarding the variability of *Ph. splendidum* from the standpoint of the genetic views prevailing at that time.

d. *Pure cultures of algae, zoochlorellae, and gonidia of lichens.*

In Part II of this book due attention has already been given to the fact that BEIJERINCK was the first to obtain pure cultures of algae, zoochlorellae, and gonidia of lichens (Cf. Chapter XVI). Although for this reason BEIJERINCK's activities in this field will not be surveyed here, it seemed desirable to include at this spot this brief reference to these studies which constitute one of the most important contributions ever made to the science of microbiology.

e. *Studies on yeasts.*

It is only natural that BEIJERINCK's industrial activities should have brought him already at the very beginning of his microbiological career to a detailed study of various yeasts. In the meantime the industrial importance of many of his investigations in this field will have prevented publication of their results. Notwithstanding this, in the course of time, BEIJERINCK was able to publish several valuable contributions to our knowledge of this group of micro-organisms.

When BEIJERINCK commenced his researches, the study of yeasts had been mainly restricted to those species and strains which found technical application in breweries, distilleries and in vinification. Following the lead of E. CHR. HANSEN, BEIJERINCK was one of the first to realize that these cultivated species were merely adapted forms of a large group of "wild yeasts" having a wide distribution in nature.

In his study on kefir BEIJERINCK gave a description of the yeast constantly present in this Caucasian product [3]). The organism had already been discovered by KERN in 1881, but BEIJERINCK added several interesting details to KERN's description. Especially note-

[1]) Folia Microbiologica **4**, 15, 1916. *Ph. splendidum* differs from the related *Ph. luminosum* by its much higher temperature optimum.

[2]) These experiments were later described in more detail by Dr. GERRETSEN. Cf. Centralbl. f. Bakt. u. Parasitenk. II, **52**, 353, 1920.

[3]) Arch. néerl. d. sciences exactes et naturelles **23**, 428, 1889.

worthy is the demonstration with the aid of the luminous-bacteria-plate method that the fermentation of the lactose is preceded by a splitting of this sugar into its hexose constituents. In addition he showed the same to be true of another lactose-fermenting yeast always found in Edam cheese, to which organism he gave the name *Saccharomyces tyrocola*.

These observations were presented in even more detail in a second paper [1]), and in it BEIJERINCK coined the term "lactase" for the enzyme which brings about the hydrolysis of lactose. Experimental proof was given that this enzyme is excreted by the yeast cells into the culture medium, and BEIJERINCK may, therefore, be rightly considered as the discoverer of lactase.

In 1892 a study [2]) was published on the nutritional requirements of the film-producing yeast species, at that time known as *Saccharomyces mycoderma*. This paper is remarkable because in it was made the first attempt to carry through a differentiation of various yeast species on the basis of their different behaviour towards sugars. Moreover, it is shown that these oxidizing yeast species are also able to develop on various organic substrates other than sugars, as for instance glycerol, succinate and acetate. This is a fact too often neglected even nowadays. Attention was also given to the suitability of various single compounds to act as sole nitrogen source for the development of various yeasts. Finally, it was emphasized, that under conditions of anaerobiosis also the oxidizing yeast is capable of bringing about a regular alcoholic fermentation; this phenomenon was discussed in the light of PASTEUR's fermentation theory.

The discovery made in 1894 of the new yeast species *Schizosaccharomyces octosporus*, isolated from raisins, may be deemed to be of great importance [3]). Here for the first time a description was given of a yeast under suitable conditions regularly producing eight endospores. This fact brought final proof for the correctness of DE BARY's and REESS' assumption that the spore-forming yeasts had to be classified with the *Ascomycetes*.

In the hands of GUILLIERMOND some years later this species was to give the first clue to the cytology and phylogeny of the whole group of yeasts. The direct inducement to these investigations may well have been BEIJERINCK's statement that — in contrast to what holds for *Saccharomyces* species — the occurrence of a nucleus in the cells of *Schizosaccharomyces octosporus* can be observed beyond any doubt. BEIJERINCK had already noted that a nuclear division into eight precedes the formation of the eight ascospores. Amongst the physiological properties of the new species, BEIJERINCK stressed its ability to fer-

---

[1]) Centralbl. f. Bakt. u. Parasitenk. **6,** 44, 1889.
[2]) Centralbl. f. Bakt. u. Parasitenk. **11,** 68, 1892.
[3]) Centralbl. f. Bakt. u. Parasitenk. **16,** 49, 1894.

ment maltose but not saccharose, a characteristic unknown for any yeast species described up to that time.

BEIJERINCK's removal to academic surroundings made him three years later decide to reinvestigate the species in question [1]). Useful indications are given for the isolation of the organism: the relatively great thermostability of the ascospores in the dry state making it possible to bring about a separation from other yeast species simultaneously present on the surface of the dried fruits from subtropical regions (raisins, figs, dates). The occurrence of asporogenous strains is dealt with in detail. Much attention is also given to gelatine liquefaction, which phenomenon is especially marked in the stage of spore formation and liberation. It remains surprising that the isogamic copulation as an introduction to the formation of the asci has escaped BEIJERINCK's attention, the more so since in the explanation to the figures he mentions that many of the asci are characterized by their yoke-like shape.

A paper of considerable methodical and theoretical interest appeared in 1898 [2]). In this memoir BEIJERINCK dealt with the difficult question of the loss of spore-forming power which is only too frequently observed in sporogenous yeast species, on continued cultivation in pure culture. BEIJERINCK found that colonies of *Schizos. octosporus* which had originated from ascospores always after some time formed ascospores again, but that this formation did not occur when the colonies were derived from ordinary vegetative cells. Upon this he based a method for regeneration, or at least for intensification, of the spore-forming power: cultures in which only rare ascospores were present were submitted to desiccation and heating at 50° C. Under these conditions the vegetative cells were as a rule killed, whilst the few ascospores present withstood this operation, so that on streaking a suspension of the dried material on wort-gelatine, colonies were obtained which formed spores abundantly. The method was applied successfully to various yeast cultures which had nearly lost their spore-forming power. Several useful indications were also given, which enabled a quick and easy differentiation between spore-forming and non-spore-forming colonies.

In the light of the present state of our knowledge regarding the life cycle of yeasts, it is noteworthy that BEIJERINCK should have described the occurrence in several cases of special strains which differed from the original culture by the much smaller dimensions of their cells. Although no interpretation of this phenomenon was offered, it seems likely that these strains must be considered as haploid forms. Such forms were later described and recognized as haploid by KRUIS and

---

[1]) Centralbl. f. Bakt. u. Parasitenk. II, **3,** 449 und 518, 1897.

[2]) Centralbl. f. Bakt. u. Parasitenk. II, **4,** 657, 1898; Arch. néerl. d. sciences exactes et naturelles Sér. II, **2,** 269, 1899.

SATAVA [1]). The independent rediscovery of these facts by WINGE [2]) has recently opened quite new and fundamentally important prospects for the study of yeasts [3]).

The next investigation in the yeast domain appeared ten years later and is of mainly physiological interest [4]). In this publication the agglutination of yeast cells was discussed. It was pointed out that there are several yeast types showing the phenomenon of auto-agglutination. Other strains, however, like the ordinary baker's yeast and the top yeasts of breweries do not have this property, but they can be induced to agglutinate by the addition of special types of lactic acid bacteria, as was first observed by BARENDRECHT [5]). Prescriptions for the identification and isolation of these bacteria were given. The paper is also of interest because it gives several details regarding the wild yeasts occurring more or less regularly at that time in commercial baker's yeasts (*Saccharomyces fragans, S. curvatus, S. muciparus, S. disporus*). Finally a method was devised for the quantitative determination of bottom yeast in a mixture with baker's yeast. This method, which depends on the specific ability of bottom yeast to attack melibiose, has not received due consideration until recently.

In a short note published in 1913 BEIJERINCK brought forward experimental proof that the then current procedure for discriminating between living and dead yeast cells with the aid of methylene blue is liable to lead to confusion if applied to yeast dried at a low temperature [6]). It was shown to be possible to obtain preparations in which all cells, though staining a deep blue on addition of the dye, still maintained their viability as could be proved by making them germinate under suitable conditions.

About the same period, a study was made of the factors determining auto-fermentation in yeast [7]). BEIJERINCK concludes from his observations that all factors which are harmful for the yeast cells lead to auto-fermentation; this point of view lured him on to some highly speculative ecological considerations.

In a joint publication with J. J. VAN HEST [8]) experiments were reported dealing with LEBEDEFF's maceration juice. The paper is mainly of interest because in it BEIJERINCK emphatically opposed the view, current at that time, that zymase was nothing but a definite chemical compound present in the dissolved state in the yeast cells and endowed with the property of splitting sugars into carbon dioxide

[1]) K. KRUIS and J. SATAVA, O Vývoji a Klíčeni Spór Jakož i Sexualitě Kvasinek. V Praze, 1918.

[2]) Ö. WINGE, C. R. Trav. Lab. d. Carlsberg, Sér. Physiol. **21**, 77, 1935.

[3]) Cf. Ö. WINGE and O. LAUSTSEN, C. R. Trav. Lab. d. Carlsberg, Sér. Physiol. **22**, 99, 1937; Ibid. **22**, 235, 1938; Ibid. **22**, 337, 1939; Ibid. **22**, 357, 1939.

[4]) Centralbl. f. Bakt. u. Parasitenk. II, **20**, 641, 1908.

[5]) H. P. BARENDRECHT, Centralbl. f. Bakt. u. Parasitenk. II, **7**, 623, 1901.

[6]) Proc. Kon. Akad. v. Wet. Amsterdam **21**, 930, 1913.

[7]) Livre jubilaire HENRI VAN LAER, p. 128, 1913.

[8]) Folia Microbiologica **4**, 107, 1916.

and alcohol. In opposition to this view, BEIJERINCK maintained that zymase is an essential, microscopically visible part of the yeast protoplasm, and therefore occurs in the maceration juice as a suspensoid. Although later investigations have more or less justified this opinion, the conclusion of the authors that zymase will never pass undamaged cell walls needs further confirmation.

BEIJERINCK's last contribution [1]) to our knowledge of the yeasts dealt with a noteworthy phenomenon, the cause of which is not yet fully understood. Many yeast species are known which owe their red colour to the presence of a pigment of carotenoid nature. BEIJERINCK now made the observation that several yeast species — as, for instance, *Saccharomyces pulcherrimus* and various yeasts isolated from milk — which under normal conditions are colourless, produce a red pigment only when grown on media containing somewhat larger quantities of iron salts. The nature of this red pigment is as yet unknown, but in any case it is not related to the carotenoids.

f. *Beijerinck's contribution to the virus concept.*

In 1898 BEIJERINCK published a paper [2]) which has since made him known as one of the pioneers in the field of virus study, so important nowadays. The paper deals with BEIJERINCK's observations on the tobacco mosaic disease. In it ample proof is afforded that the contagious agent causing the disease does not belong to the visible microorganisms, but on the contrary is a principle which occurs in the plant juice in a "dissolved state", *i.e.*, passes filters which retain all microscopically visible particles.

In the introduction to the paper BEIJERINCK states the reasons which led him to his investigation. They seem sufficiently interesting to report them briefly here. In 1885 while he was still working in the Agricultural College at Wageningen, his colleague ADOLF MAYER brought experimental proof for the contagious character of the mosaic disease. At MAYER's request, BEIJERINCK made an attempt to isolate the responsible micro-organism, but the result of his investigation was entirely negative. However, on account of the very restricted bacteriological experience which he possessed at that time, BEIJERINCK himself did not consider this result to be conclusive. The successful isolation of the root nodule organism in 1887 encouraged him to make another attempt to isolate the causative organism of mosaic disease in tobacco. The consideration that it was not the special task of an industrial microbiologist to solve the riddle of a contagious plant disease does not have seem weighed heavily with BEIJERINCK. This

---

[1]) Arch. néerl. d. physiol. de l'homme et des animaux **2,** 609, 1918.

[2]) A preliminary publication appeared in Versl. Kon. Akad. v. Wet. Amsterdam **7,** 229, 1898. A more extensive memoir was published in Arch. néerl. d. sciences exactes et naturelles Sér. II, **3,** 164, 1900.

consideration may nevertheless have been responsible for the fact that evidently he soon resigned himself to another negative result. After having become an academic teacher, he felt quite free in the choice of the subject of his researches, and, since the opening of the Bacteriological Institute had provided him with all means necessary for the investigation in question, he returned in 1897 to the problem offered by tobacco mosaic.

This time he was able to provide definite proof that the juice obtained by expressing the leaves of diseased plants contained a principle which passed through a porcelain filter retaining all visible microorganisms, which principle on being inoculated into a healthy tobacco plant, transmitted the disease to it.

Moreover, it was demonstrated that the principle actually multiplied in the living tissues of the host, so that infection in series could be obtained. In addition it was shown that the principle shared with most living cells the property of being destroyed by heating the juice to 90° C. Great stress was laid by BEIJERINCK on the outcome of the experiment in which he proved that on bringing a drop of the juice of diseased plants on the surface of an agar gel the contagious principle diffused into this gel, so that after a week or ten days its presence could be demonstrated in a layer at least two millimeters beneath the surface. For BEIJERINCK this meant a convincing proof of the noncorpuscular nature of the principle, which, therefore, should occur in the liquid state in the juice [1]). This led him to the characterization of the principle as a *"contagium vivum fluidum"* [2]). When to the foregoing we add that BEIJERINCK also proved that the contagium multiplied only in tissue in which cell division took place, and that, moreover, it could be dried at low temperature or precipitated with alcohol from the aqueous solution without loss of infectivity, it will be obvious that he succeeded in establishing the main properties characteristic for all viruses.

The great merit of this pioneer investigation in the virus field is not diminished by the fact that shortly after the appearance of the preliminary communication a note was published by IWANOWSKI [3]) in which this author rightly claimed the priority for the discovery of the filtrability of the contagious agent of mosaic disease. In a paper which had been published already seven years before IWANOWSKI had indeed proved this fact beyond doubt [4]). BEIJERINCK, to whom this publication had remained unknown, readily acknowledged this claim both in a separate note [5]), and in an addendum to the French version of his extensive publication [6]).

[1]) The expressions "liquid state" and "dissolved state" of the virus were apparently employed by BEIJERINCK interchangeably.

[2]) It is, however, noteworthy that BEIJERINCK uses this indication only in the title of the paper, but not in the text, wherein the term "virus" is used throughout.

[3]) D. IWANOWSKI, Centralbl. f. Bakt. u. Parasitenk. II, **5**, 250, 1899.

[4]) D. IWANOWSKI, Bull. de l'Acad. Imp. d. St. Pétersbourg **13**, 237, 1892.

[5]) Centralbl. f. Bakt. u. Parasitenk. II, **5**, 310, 1899.

[6]) Cf. footnote 2 on page 116.

But anybody reading IWANOWSKI's 1899 paper will have to acknowledge that this author, even seven years after he made his discovery, was not at all aware of its tremendously far-reaching importance, the main part of the paper being devoted to an attempt to prove contrary to all available evidence the bacterial nature of the contagious agent.

In contrast to IWANOWSKI's attitude, BEIJERINCK expresses throughout his paper a firm belief in the existence of an autonomous sub-microscopical form of life, and he also stresses his conviction that the case of the mosaic disease will not stand alone. In a final paragraph he mentions several instances of plant diseases which might equally be due to a "contagium fluidum", and it is clear that he already foresaw the great significance which virus diseases would acquire in phytopathology.

In this first paper, BEIJERINCK did not give much attention to the consequence of his findings from the standpoint of general biology. However, he enlarged on this point in 1913 in the very attractive address he delivered in the joint meeting of the sections of the Koninklijke Akademie van Wetenschappen at Amsterdam [1]). In this address which bore the title "Infusions and the discovery of bacteria" he dealt with the question of submicroscopical life in an eloquent way, as may be judged from the following translation of his concluding remarks:

"The existence of these contagia proves that the concept of life — if one considers metabolism and proliferation as its essential characters — is not inseparably linked up with that of structure; the criteria of life, as we find it in its most primitive form, are also compatible with the fluid state."

And somewhat further on:

"In its most primitive form, life is, therefore, no longer bound to the cell, the cell which possesses structure and which can be compared to a complex wheel-work, such as a watch which ceases to exist if it is stamped down in a mortar.

"No; in its primitive form life is like fire, like a flame borne by the living substance; — like a flame which appears in endless diversity and yet has specificity within it; — which can adopt the forms of the organic world, of the lank grass-leaf and of the stem of the tree; — which can be large and which can be small: a molecule can be aflame; — which can be so nearly luke warm as not to scorch the human hand; — which is bound up with a material foundation and yet leads to immaterial consequences; — which yields energy and converts energy into other forms; — which acts as a catalyst that brings about in its environment changes all out of proportion to its own size; — which consumes oxygen and excretes carbon dioxide; — which ab-

---

[1]) Jaarboek der Koninklijke Akademie van Wetenschappen voor 1913.

sorbs nutrients, can multiply itself and divide; — which does not originate by spontaneous generation, but is propagated by another flame."

However vague these thoughts may be, yet they seem to justify the eulogy which another great microbiologist, FÉLIX D'HÉRELLE, pronounced twelve years later in the Amsterdam Academy:

"On a beaucoup discuté la conception de BEIJERINCK, mais je ne pense pas qu'on en ait saisi toute la profondeur. Toute la biologie reposait, repose encore, sur l'hypothèse fondamentale que l'unité de matière vivante, c'est la cellule. BEIJERINCK le premier, s'est affranchi de ce dogme, et a proclamé de fait, que la vie n'est pas le résultat d'une organisation cellulaire, mais dérive d'un autre phénomène, qui ne peut dès lors résider que dans la constitution physico-chimique d'une micelle protéique." [1]).

Those who have been watching the recent developments in the study of the viruses, especially the developments arising from STANLEY's great discovery of crystalline mosaic virus, will commend the appositeness of the consideration formulated by D'HÉRELLE. It even may be expected that thoughts like these are bound to play an important rôle in the further elucidation of the phenomenon of life.

That BEIJERINCK in his later years retained his concern with the problems of submicroscopical life may be inferred from the fact that he published an essay on "Pasteur and ultramicrobiology" in 1922, on the occasion of the centenary of PASTEUR's birthday [2]).

g. *Investigations on lactic acid bacteria.*

One of the chief contributions of BEIJERINCK to general bacteriology has been his early recognition of the existence of the natural group of true lactic acid bacteria. At the time that BEIJERINCK entered the bacteriological field, and for many years after, there was still a strong tendency to consider any bacterium as a lactic acid bacterium, if under certain conditions it produced lactic acid from sugars. BEIJERINCK's work has done much to promote the view that the term "lactic acid bacterium" should be restricted to representatives of a natural group of bacteria, which, in addition to their property of producing lactic acid, have many other characteristics in common. It should be added, however, that it was only the appearance in 1919 of ORLA-JENSEN's monograph "The Lactic Acid Bacteria" that brought finality to the discussion.

BEIJERINCK's occupations with the lactic acid bacteria had a twofold origin. In the first place, his activity in the fermentation industry forced him to give full attention to the various types of lactic acid bacteria which play either a desirable or an undesirable rôle in the

[1]) Versl. Afd. Natuurk. Kon. Akad. v. Wet. Amsterdam **34,** 835, 1925.
[2]) Chemisch Weekblad **19,** 525, 1922.

commercial production of yeast. Secondly the gradual introduction of scientific principles into the dairy industry led to an increased interest in the bacteriological processes which are the basis of butter and cheese manufacture.

Moreover, at the outset of BEIJERINCK's bacteriological career, various sour milk preparations, like kefir and yoghurt, were introduced into Western-Europe, and it is only natural that a bacteriologist with so wide an interest as BEIJERINCK had, should wish to take part in the investigation of the agents which are active in the preparation of such products. The first of BEIJERINCK's papers to deal in detail with a lactic acid bacterium referred to the microbiology of sour milk preparations. In 1889 he published a study on kefir which can be characterized even to-day as the most outstanding contribution to our knowledge of this remarkable "milk-ferment" [1]). BEIJERINCK gave convincing evidence that the kefir grains consist of zoogloea of a lactic acid bacterium (*Bacillus caucasicus*) surrounded by layers of cells of a special lactose-fermenting yeast, *Saccharomyces kefyr*. The Russian investigator KERN who in 1882 had given a rather confusing description of the micro-organisms present in kefir had proposed the name of *Dispora caucasica* for the bacterial constituent, since he thought he had seen the formation of two spores in each cell. It remained for BEIJERINCK to prove that the organism in question was a typical non-spore-forming lactic acid bacterium. Characteristic of BEIJERINCK's broad views is his emphasis on the fact that symbiosis between yeasts and lactic acid bacteria is not at all restricted to kefir, but is on the contrary, of quite general occurrence. He cited the examples of Edam cheese, ensilage, leaven, the faeces of breast-fed infants, and the applications made in the fermentation industries.

The detection of lactic acid bacteria was greatly facilitated by BEIJERINCK's suggestion either to cover suitable agar and gelatine media with a thin layer of finely divided chalk, or to incorporate this material into such media [2]).

Acid-producing colonies are then quickly surrounded by clear zones which contrast markedly with the rest of the opaque plates.

In a very short note published in 1893 BEIJERINCK reported the rather startling observation that lactic acid bacteria were exceptions to the rule universally accepted at that time, that every living cell has the ability to decompose hydrogen peroxide into water and free oxygen [3]). In the Delft school of bacteriology this observation has since been always applied for diagnostic purposes. For bacteria growing under aerobic conditions the easily-established finding "catalase-negative" practically always justifies the diagnosis of "lactic acid bacterium". Only a very few exceptions to this rule have been found in

---

1) Arch. néerl. d. sciences exactes et naturelles **23,** 428, 1889.
2) Centralbl. f. Bakt. u. Parasitenk. **9,** 781, 1891.
3) Naturw. Rundschau **8,** 671, 1893.

the 45 years which have passed since BEIJERINCK made the observation referred to above [1]).

It was not until 1901 that BEIJERINCK decided to make public the vast experience which he had gathered during his industrial period regarding the lactic acid bacteria active in the yeast and alcohol industry [2]). In this publication a survey is given of the various rod-shaped lactic acid bacteria which are frequently encountered in the industry in question. BEIJERINCK laid it down that they constitute a group which is homogeneous both from the morphological and from the physiological point of view, and accordingly he felt justified in ascribing to this group the natural rank of a genus, for which he proposed the name *Lactobacillus*.

For a survey of the various *Lactobacillus* species with which BEIJERINCK had become acquainted in his industrial period, it may suffice to refer the reader to the original paper, and to the thesis of JAN SMIT which thesis about ten years later was prepared under BEIJERINCK's supervision [3]). It seems worth-while, however, to mention here briefly the various properties which BEIJERINCK considered to be characteristic for true lactic acid bacteria.

In the first place BEIJERINCK stressed the absence of hydrogen in the fermentation gas, when such gas is produced. BEIJERINCK pointed out that this characteristic is of significance in the differentiation of the lactic acid bacteria from the bacteria belonging to the genus *Aerobacter*, as outlined by him [4]), which bacteria also produce larger or smaller quantities of lactic acid from sugar. There seems little doubt that the criterion in question is quite valid, though it has remained unnoticed by later investigators. Even in ORLA-JENSEN's classical monograph, "The Lactic Acid Bacteria", which appeared in 1919, one finds the casual remark that hydrogen may occur in the fermentation gas produced by true lactic acid bacteria. Since, however, no documentation for this contention is presented, ORLA-JENSEN's remark may be considered as a relic of the confusion which formerly existed regarding the definition of "lactic acid bacteria".

Further general characteristics of true lactic acid bacteria as sustained by BEIJERINCK are: complete immotility in all stages of development; the small dimensions of the colonies, even under favourable nutritional conditions; and the absence of catalase, as already discussed above. In addition it is pointed out that peptones are the only suitable nitrogen source for the lactic acid bacteria. This

---

[1]) For an acetic acid bacterium not containing catalase, *viz.*, *Acetobacter peroxydans*, cf. F. VISSER 'T HOOFT, Biochemische onderzoekingen over het geslacht *Acetobacter*, Delft 1925, and also H. WIELAND und H.J.PISTOR, Ann. d. Chemie **522,** 116, 1936.

[2]) Arch. néerl. d. sciences exactes et naturelles Sér. II, **6,** 212, 1901.

[3]) JAN SMIT, Bacteriologische en chemische onderzoekingen over de melkzuurgisting. Diss. Amsterdam, 1913.

[4]) It should be realized that nowadays the genus *Aerobacter* BEIJERINCK is used in a much more restricted sense. As originally created by BEIJERINCK it was meant to embrace all bacteria of the so-called coli-aerogenes group.

view was generally accepted by all investigators in this field until ORLA-JENSEN recently proved that ammonium nitrogen is also assimilated, provided that suitable activators of an organic nature are present.

Finally, BEIJERINCK expresses the opinion that the production of mannitol from laevulose is another general property of lactic acid bacteria. Nowadays we know that this ability is restricted to the subgroup of the so-called heterofermentative lactic acid bacteria. Yet it must be deemed to be a great merit of BEIJERINCK that he fully realized that the said mannitol production is not due to the action of a separate group of specific bacteria, as is suggested by the term "ferment mannitique", often used by French investigators, even nowadays.

Six years later BEIJERINCK published another fundamentally important paper on the group under review, this time dealing with the various lactic acid bacteria occurring in milk and in milk products [1]).

This presented an even more systematic survey of the properties of the true lactic acid bacteria. In addition to the characters already discussed BEIJERINCK laid emphasis upon the absence of spore-forming ability. He also pointed out, however, that the cells of the lactic acid bacteria are much more thermo-resistant than those of other non-spore-forming bacteria, so that by heating to 65–75° C. during a suitable period a separation of the lactic acid bacteria from the other non-spore-forming groups can be carried through. For this procedure he coins the new term "lacticisation", a term which, however, has not found acceptance.

BEIJERINCK also expressed the opinion that all lactic acid bacteria have the ability to decompose the glucosides aesculin and indican (emulsin reaction). For the detection of this property he devised very simple and elegant methods [2]). In the light of our present-day knowledge it seems doubtful whether, indeed, the said property is a universal characteristic of lactic acid bacteria. It seems more probable that the reaction is restricted to the homofermentative subgroup, as also holds for the decomposition of another glucoside, salicin.

The greater part of the paper is devoted to a detailed description of the different types of lactic acid bacteria which are obtained in accumulation experiments with milk at different temperature ranges. A final chapter deals exhaustively with the lactic acid bacteria present in yoghurt, a preparation which at that time had just come into vogue owing to METCHNIKOFF's suggestive theory of auto-intoxication caused by the normal intestinal flora.

---

1) Proc. Kon. Akad. v. Wet. Amsterdam **10,** 17, 1907.

2) As far as aesculin decomposition is concerned the observations were due to BEIJERINCK's colleague TER MEULEN, who had previously made extensive studies on glucoside decomposition.

h. *Investigations on the natural group of butyric acid and butyl alcohol bacteria.*

A second natural group of bacteria which became the subject of a thorough investigation by BEIJERINCK is that of the anaerobic, spore-forming, sugar-fermenting bacteria, generally known by their common names of butyric acid and butyl alcohol bacteria.

BEIJERINCK's first publication on the bacteria of this group was of an astonishingly exhaustive character [1]). Judging from the title, "Sur la fermentation et le ferment butyliques", one might expect that this publication would be restricted to the butyl alcohol fermentation and its causative organism. In reality, however, BEIJERINCK gave a critical survey of the whole group of anaerobic, spore-forming, sugar-fermenting bacteria. The discussion is obviously based on extensive personal experiences with the main representatives of this group.

The introduction opens with the remark that the author already in 1886, *i.e.*, seven years before, had found that certain cereal grains — more especially barley — after having been ground and soaked with boiling water, readily enter into a gassy fermentation, amongst the products of which butyl alcohol is easily detected. Starting from other samples, however, in many cases butyric acid is the most characteristic fermentation product.

At first sight it is a little surprising that BEIJERINCK postponed the publication of his studies on the butyl alcohol fermentation so long. His statement that in 1885, after the death of FITZ, he received the strains of *Bacillus butylicus* described some years earlier by this pioneer in the fermentation field, supplies evidence that BEIJERINCK had already thoroughly studied that bacterial group at the very beginning of his microbiological career.

Obviously, at that time, other problems took his attention before he found the time necessary for concluding his investigations by a publication.

In the introductory remarks to his 1896 paper BEIJERINCK emphasizes that normal butyl alcohol frequently occurs in the fermentation of sugars by various bacterial species. Yet he added that, as a rule, this alcohol has only the character of a minor product accompanying larger amounts of butyric acid. This holds, for example, for the fermentation caused by *Bacillus butylicus* Fitz. BEIJERINCK also referred to the fermentation caused by GRIMBERT's *Bacillus orthobutylicus* as being of the butyric acid type. BEIJERINCK, therefore, expressed as his opinion that the butyl alcohol fermentation described by him in such profuse detail differs in principle from all so-called "butyl alcohol fermentations" reported up to that time.

---

[1]) Verhandelingen Kon. Akad. v. Wet. Amsterdam, 2de Sectie, **1,** No. 10, 1893. A French, somewhat extended, version of this memoir was published three years later Arch. néerl. d. sciences exactes et naturelles **29,** 1, 1896.

Although this statement is perhaps not fully justified as regards GRIMBERT's results [1]), it cannot be denied that BEIJERINCK's paper meant considerable progress; it may even be regarded as the foundation stone of our knowledge of a fermentation process which in recent years has obtained such a considerable economic importance [2]). BEIJERINCK's contribution is chiefly of importance, because it supplies detailed prescriptions for isolation of the causative organism with the aid of well-devised enrichment experiments. Another valuable point is the recognition of the close relationship between the "butyl ferment" and the other spore-forming sugar-fermenting bacteria of which two anaerobic, butyric acid forming, types are described together with PRAZMOWSKI's facultatively anaerobic species, *Bacillus polymyxa*. All these species were united by BEIJERINCK into one genus for which the name *Granulobacter* was proposed on the ground of the common property that under certain conditions the cells take the form of clostridia staining blue on addition of iodine, due to the presence of a reserve carbohydrate, to which the name granulose was given. Besides the diagnosis of the genus, BEIJERINCK gave a full description of the four *Granulobacter* species with which he had become intimately acquainted.

Special attention may be called to BEIJERINCK's intuition which made him at once discriminate between the sugar- and the lactate-fermenting butyric acid bacteria. The recent work of VAN BEYNUM and PETTE [3]) has thrown full light on the great practical importance of this differentiation.

The greater part of BEIJERINCK's paper supplies an exemplary description of his butyl ferment, *Granulobacter butylicum*. Both its morphological and its physiological characteristics are dealt with in great detail. Whilst BEIJERINCK thought that besides the butyl alcohol, normal propyl alcohol was produced [4]), it has since been established with certainty that the organism in question produces isopropyl alcohol. In 1929 VAN DER LEK [5]) revived BEIJERINCK's organism from an old dried spore culture labelled by BEIJERINCK in 1893, the bacterium having remained viable in its resting stage for at least 36 years! VAN DER LEK then made accurate determinations of all fermentation products and found that isopropyl, and not normal propyl, alcohol was always present in considerable amounts in the neutral volatile fraction. He thereby offered definite evidence that BEIJE-

[1]) L. GRIMBERT, Ann. de l'Inst. Pasteur **7**, 353, 1893.

[2]) Significant in this respect is BEIJERINCK's remark that, if butyl alcohol were a product of technical importance, it could easily and cheaply be prepared by the fermentation method.

[3]) J. VAN BEYNUM und J. W. PETTE, Zentralbl. f. Bakt. II, **93**, 198, 1935; Ibid. **94**, 413, 1936.

[4]) In a later paper (cf. Proc. Kon. Akad. v. Wet. Amsterdam **1**, 14, 1898) BEIJERINCK even goes as far as to say that his organism produces much more propyl alcohol than butyl alcohol and suggests that, therefore, the name *Granulobacter propylicum* would be more appropriate. This, evidently, is a lapsus.

[5]) J. B. VAN DER LEK, Onderzoekingen over de butylalkoholgisting. Delft, 1930.

RINCK's butyl ferment is not identical with *Clostridium acetobutylicum*, the organism nowadays generally used in the technical production of butyl alcohol. In the fermentation caused by the latter species, not isopropyl alcohol, but the closely related acetone, occurs.

Apart from a short notice on an enrichment procedure for his butyric acid bacterium, *Granulobacter saccharobutyricum* [1]), BEIJERINCK devoted only one more publication to the representatives of the genus in question. This study, made jointly with his assistant A. H. VAN DELDEN, dealt more especially with the bacteria active in the retting of flax [2]).

In May 1903 a committee had been appointed, charged with the task of investigating the possibility of promoting the home working up of flax grown in the Netherlands. Until that time, by far the greater part of the flax harvested in the northern provinces of the Netherlands (Friesland and Groningen) was sent to Belgium and submitted to a retting process in the river Lys near Courtrai. BEIJERINCK accepted the task of studying the applicability of the warm water retting process, introduced into Belgium some years before.

In BEIJERINCK and VAN DELDEN's study the fundamentals of the retting process are clearly exposed. From an anatomical study of the flax plant, convincing evidence was derived that retting is essentially a process of pectin fermentation which liberates the fibres from the surrounding parenchyma and the central woody stem. It is pointed out that a satisfactory retting procedure depends on a successful enrichment culture of pectin-fermenting bacteria. It is then shown that at least under the chosen conditions of warm water retting, pectin fermentation is due to the action of a plectridium-forming *Granulobacter* species, to which the name of *Gr. pectinovorum* is given [3]). This bacterium which, in contrast to the other species of the genus, also readily ferments pectin in synthetic media, is apparently identical with the *Plectridium pectinovorum* described a year before by STÖRMER [4]). Yet, BEIJERINCK's careful observations added a good deal to our knowledge of the organism. Besides this principal agent of the retting process another new species, *viz.*, *Granulobacter urocephalum* was encountered, and a description of this species was given, though it is apparently only of secondary importance in the retting process.

Finally mention should here be made of a study published by BEIJERINCK, jointly with his collaborator DEN DOOREN DE JONG, at the end of his scientific career, *i.e.*, shortly after his retirement from the chair at Delft [5]). The paper bears the short title "On *Bacillus polymyxa*" and deals with the remarkable bacterium already described

[1]) Centralbl. f. Bakt. u. Parasitenk. II, **2**, 699, 1896.

[2]) Arch. néerl. d. sciences exactes et naturelles Sér. II, **9**, 418, 1904.

[3]) Cf., however, A. D. ORLA-JENSEN und A. J. KLUYVER, Zentralbl. f. Bakt. II, **101**, 1939.

[4]) K. STÖRMER, Mitt. d. deutschen landwirtsch. Gesellschaft p. 193, 1903. Cf. abstract in Centralbl. f. Bakt. u. Parasitenk. II, **11**, 66, 1904.

[5]) Proc. Kon. Akad. v. Wet. Amsterdam **25**, 279, 1922.

by PRAZMOWSKI in 1880, which BEIJERINCK, in 1893, had also incorporated into his genus *Granulobacter*. *Bacillus polymyxa* has in common with the other *Granulobacter* species that it is a spore-forming rod which brings about a vigorous sugar fermentation. However, *Bacillus polymyxa* occupies a very special position, because, unlike the obligatory anaerobic butyric acid and the butyl alcohol bacteria, it attains a good development under aerobic conditions and, therefore, forms a transition to the aerobic, non-fermenting, spore-forming bacteria of the genus *Bacillus* [1]).

The paper affords several points of interest. In the first place it throws much light on the wide occurrence of *Bacillus polymyxa* in nature, and it describes also suitable enrichment procedures facilitating its isolation. The main point of interest, however, is the study of the conditions which determine the production of large amounts of mucus by the bacterium, a property which, having already been observed by PRAZMOWSKI, was responsible for the specific name of the organism. BEIJERINCK and DEN DOOREN DE JONG proved that this production of a slime of carbohydrate nature is due to a lack of balance in the nutritional factors: a shortage of assimilable nitrogenous substances in presence of an excess of carbohydrate in the medium being especially favourable for the formation of the mucus. The demonstration that the mucus, once formed, is consumed again by the bacterium itself on addition of a nitrogenous compound, like asparagine, and is then converted into new cell material, is particularly convincing in this respect. There is no doubt that the same situation obtains with many slime-producing bacteria. Several recent studies on bacteria of this type might have been influenced for the better, if the investigators performing them had been acquainted with the fundamentally important observations referred to above.

i. *The genus Aerobacter Beijerinck.*

After what has been reported on BEIJERINCK's work on the lactic acid and the butyric acid bacteria, it seems only natural for BEIJERINCK to have given due attention to the third important natural group of sugar-fermenting bacteria, frequently known by its vulgar name of "colon group". Yet, BEIJERINCK's communications regarding his experiences with the representatives of this group are chiefly confined to one publication which first appeared in 1900 [2]). The title of the paper, "Sur la formation de l'hydrogène sulfuré dans les canaux, et le genre nouveau *Aerobacter*", would make one conclude that

[1]) The situation outlined above has made DONKER decide to unite *Bacillus polymyxa* together with a few other related species into a new genus for which the name *Aerobacillus* was proposed. Cf. H. J. L. DONKER, Bijdrage tot de kennis der boterzuur-, butylalcohol- en acetongistingen. Delft, 1926.

[2]) Centralbl. f. Bakt. u. Parasitenk. II, **6**, 193, 1900. A somewhat extended version was published in: Arch. néerl. d. sciences exactes et naturelles Sér. II, **4**, 7, 1901.

BEIJERINCK's treatment of the group was merely a by-product of his studies on the formation of hydrogen sulphide in nature. A scrutiny of the chapter entitled "Creation of the genus *Aerobacter*" leaves no doubt that BEIJERINCK had made himself quite generally acquainted with the various species belonging to the new genus.

As will be seen in one of the following sections, BEIJERINCK had at that time already made important observations regarding the process of sulphate reduction, the chief biological source of hydrogen sulphide in nature. However, BEIJERINCK emphasized from the very beginning that there are also minor biological sources of hydrogen sulphide, and he early indicated the bacterial decomposition of sulphur-containing proteins as one of these.

The regular production of bad smells by the water of the canals in Delft during the hot summer-time made BEIJERINCK decide to investigate whether a decomposition of sulphate, accompanied by formation of hydrogen sulphide, also took place under the semi-aerobic conditions prevailing in the canal-water. By dispersing finely-divided lead carbonate in ordinary nutrient peptone agar he was able to obtain white media on which many bacteria present in the water appeared to develop quite satisfactorily. The great advantage of this medium is that all bacteria, producing hydrogen sulphide from the peptone, can be recognized at a glance because of the brown to black colour conferred on the colonies by the formation of lead sulphide.

BEIJERINCK found that the majority of the sulphide-forming colonies could be identified with one of the two bacterial species then called *Bacterium coli commune* and *Bacterium lactis aerogenes*.

Numerous publications dealing chiefly with the hygienic significance of these bacteria had already appeared, and from these it had become clear that there exist many varieties and intermediate forms of these two species. On studying the group, BEIJERINCK almost completely ignored the hygienic questions. His first impulse was characteristically scientific: namely, to stress the desirability of separating the two species and their related forms from the many other forms which at the time were designated by the generic name *Bacterium*. He, therefore, proposed the new genus *Aerobacter* to contain the species under consideration. From the diagnostic characters of the genus laid down by BEIJERINCK, the following may be mentioned: non-spore-forming rods, either motile or non-motile, which ferment various sugars and poly-alcohols with production of lactic acid and gas, the latter always consisting of a mixture of hydrogen and carbon dioxide; nitrates are easily reduced with formation of nitrites, but sulphates are not attacked.

BEIJERINCK added some interesting remarks on the relation of his new genus to other natural groups, especially to the aerobic spore-forming bacteria (for which, incidentally, he introduced the two new generic names *Fenobacter* and *Saccharobacter*) and to the groups of the butyric acid and of the lactic acid bacteria.

Within the genus *Aerobacter* BEIJERINCK recognized four different species [1]). For the first two species, *viz.*, *A. aerogenes* and *A. viscosum*, which may be easily distinguished on the ground of the strong slime-forming ability of the latter, a simple enrichment procedure is described. The third species *Aerobacter coli* is the typical organism predominating in human faeces. In this species BEIJERINCK created the var. *infusionum*, for a form frequent in plant infusions and in water polluted with vegetable remains.

The fourth species, *Aerobacter liquefaciens*, is worthy of some special attention. It differs from the foregoing species by its ability to bring about marked liquefaction of gelatine. Another characteristic of this species did not escape BEIJERINCK's attention, *viz.*, the fact that the cells of this species are motile with the aid of one polar flagellum, in contrast to what holds for the other species, strains of which, if motile, have peritrichous flagella.

At that time this difference was judged to be not incompatible with generic identity. In later years, such a difference has usually been deemed to be of decisive significance for the separation of natural groups. It was mainly from the type of flagellation that VON WOLZOGEN KÜHR decided to incorporate the fermenting organism with polar flagella, isolated by him, into the genus *Pseudomonas* [2]). There seems, nowadays, little doubt that the bacterium described by KÜHR as the new species *Pseudomonas fermentans* is identical with BEIJERINCK's *Aerobacter liquefaciens.*

On concluding this survey of the genus *Aerobacter*, as outlined by BEIJERINCK, it seems necessary, in order to avoid misunderstanding, to observe that the generic name *Aerobacter* Beijerinck is used in a much more restricted sense in the well-known American classification of BERGEY *et al.*

In this classification the genus embraces only those species which produce acetyl methyl carbinol in the sugar fermentation and, therefore, give a positive VOGES and PROSKAUER reaction [3]).

j. *Investigations on Sarcina ventriculi.*

Whilst the fermentation processes mentioned in the preceding paragraphs have been known from the very beginning of the development of bacteriology as a science, it was also given to BEIJERINCK to

---

[1]) For the description of these species one should also compare BEIJERINCK's paper on indigo fermentation in Proc. Kon. Akad. v. Wet. Amsterdam **2,** 495, 1900.

[2]) C. A. H. VON WOLZOGEN KÜHR, Zentralbl. f. Bakt. II, **85,** 223, 1932. Recently KLUYVER and VAN NIEL have gone so far as to create a new genus: *Aeromonas* for the *Pseudomonas* species having the type of fermentation characteristic of *Ps. fermentans.* Cf. A. J. KLUYVER and C. B. VAN NIEL, Zentralbl. f. Bakt. II, **94,** 369, 1936.

[3]) Cf., however, A. J. KLUYVER and E. L. MOLT, Proc. Kon. Ned. Akad. v. Wet. Amsterdam **42,** 118, 1939.

discover in 1905 a fermentation process which had remained unnoticed. In a paper published in that year BEIJERINCK described an extremely interesting enrichment procedure which with almost unfailing regularity brings to the fore a large sarcina-shaped microorganism causing a vigorous fermentation in sugar containing media, such as beer wort [1]). The discovery of this quite unexpected fermentation was the result of a series of systematic experiments made — in part jointly with Dr. N. GOSLINGS — to examine the question as to which are the organisms able to develop in media of high acidity under anaerobic conditions. In this investigation it was found that if the development of moulds and yeasts was suppressed by complete exclusion of air, the addition of somewhat higher amounts of inorganic acids to beer wort inoculated with garden soil almost invariably led to a fermentation which was marked by the development of large sarcina packets.

It happened that SURINGAR, professor of botany at the University of Leiden, who had been BEIJERINCK's teacher in his student period, had published in 1865 a monograph on the remarkable sarcina noted by GOODSIR, a Scottish physician, as long ago as 1842.

GOODSIR had observed the occurrence of regularly formed packets in the stomach contents of a patient, and had described these formations under the name of *Sarcina ventriculi*. This observation was repeated from time to time by medical investigators, who encountered the organism especially in cases of *stenosis oesophagi*. It was soon suspected that a close connection might exist between the presence of the sarcinae and a gas development sometimes occurring in the stomach. However, no proof for the correctness of this assumption could be furnished, since it appeared impossible to cultivate the organism *in vitro*. SURINGAR was the first to prove the vegetable nature of the organism, and, from his time on, it has been ranked with the bacteria.

There is no doubt that BEIJERINCK was thoroughly acquainted with the organism to which his former teacher had once devoted so much of his attention. It is, therefore, not surprising that BEIJERINCK should have taken into consideration in his first paper, the possible identity of his new fermentation organism and GOODSIR's *Sarcina ventriculi*. It should, however, be realized how daring a thought this was. On the one hand an organism which appeared, on the evidence of enrichment cultures to be practically ubiquitous in nature, on the other hand a medical "living curiosity" which nobody had ever seen develop outside the human body.

BEIJERINCK's studies of his new fermentation organism had made him familiar with one especially remarkable property, *viz.*, that the cultures could only be transferred into fresh media as long as the fermentation was still active. Obviously the bacterium dies off very

[1]) Proc. Kon. Akad. v. Wet. Amsterdam **7**, 580, 1905.

quickly after fermentation ceases, partly because as a strict anaerobe it cannot withstand traces of oxygen diffusing into the medium, partly perhaps owing to the action of the organic acids formed in the fermentation.

This observation made BEIJERINCK realize that a cultivation of the stomach sarcina *in vitro* would only succeed if the stomach contents in which it was present were transferred immediately after their collection into a medium permitting optimal development. Neglect of this point might well be responsible for the failure of earlier investigators to cultivate the organism.

It was only six years later that BEIJERINCK got the opportunity to submit his hypothesis to an experimental test [1]). This test led to a completely satisfactory result. The bottles of beer wort inoculated with the fresh stomach contents of a patient entered quickly into a strong fermentation, and the causative organism could be transferred in exactly the same way as the soil organism. In other respects also complete identity of the two organisms was established.

The excellent monograph which BEIJERINCK's former collaborator SMIT in recent years has devoted to *Sarcina ventriculi* and some related organisms, throws a clear light on the remarkable properties of the representatives of this group [2]). SMIT stresses that the wide distribution of *Sarcina ventriculi* in nature seems quite opposed to the extreme sensitivity of the organism when cultivated in pure culture. A resolution of this paradox has not yet been reached. Further work on this subject seems most desirable, and may be of great importance for our general insight into the conditions which determine the survival of microbes in nature.

Finally, it seems probable that the recent procedures for the preparation and preservation of ensilage, based on the reputed absence of microbial life under anaerobic conditions as soon as the acidity of the medium corresponds to pH 4.0 or lower, may before long lead also to the realisation of the great practical significance of the fermentation process discovered by BEIJERINCK [3]).

k. *Investigations on acetic acid bacteria.*

The frequent occurrence of acetic acid bacteria in fermentation industries leaves no doubt that already very early in his career BEIJERINCK became thoroughly familiar with various types of acetic acid bacteria. Yet, it was not until 1898 that he decided to deal in a publication [4]) with his experiences on this natural group of bacteria. The reason for this decision was the circumstance that at the same time a

[1]) Proc. Kon. Akad. v. Wet. Amsterdam **13,** 1237, 1911.
[2]) JAN SMIT, Die Gärungssarcinen. Eine Monographie. Jena, 1930.
[3]) SMIT's experiments have shown definitely that development of *Sarcina ventriculi* is possible in media having a pH only slightly above 1.1.
[4]) Centralbl. f. Bakt. u. Parasitenk. II, **4,** 209, 1898.

substantial treatise on the acetic acid bacteria was published by HOYER who had been working on this subject under BEIJERINCK's supervision [1]).

Both HOYER's and BEIJERINCK's publications have, as central feature, the idea that the various acetic acid bacteria constitute a natural group, and should, therefore, be sharply differentiated from the numerous other sporeless, rod-shaped bacteria which also have an oxidative metabolism. In this respect the ability of the acetic acid bacteria to produce in suitable media high amounts of acid is a decisive characteristic; this property is accompanied by an adaptation to life in acid culture media. For this reason it is surprizing that neither BEIJERINCK nor HOYER proposed in their publications the creation of a new genus for the acetic acid bacteria. At least they neglected to do so formally, but there is sufficient evidence that soon afterwards BEIJERINCK introduced the generic name *Acetobacter* [2]) into his conversations and private correspondence. In various papers which appeared shortly after 1898, the name *Acetobacter* is used without any further explanation [3]). There can be no doubt that in any case morally, but probably also according to the letter of the code of Botanical Nomenclature, BEIJERINCK is to be considered as the author of the genus *Acetobacter*, as it occurs in most of the recent bacterial systems.

Another characteristic element in both BEIJERINCK's and HOYER's publications was the tendency to restrict as much as possible the number of the species to be distinguished within the group. Both authors were fully aware that a systematic study of the group leads to the isolation of numerous non-identical strains, but since these differences are often limited to characters of minor importance, the authors emphasized the necessity of distinguishing only a small number of species which may then each embrace a certain number of varieties. A more general application of this principle in bacterial classification would have saved this science from much confusion. For the acetic acid bacteria the result was that only four species — *B. aceti*, *B. rancens*, *B. Pasteurianum* and *B. xylinum* — were recognized and clear differential characters were given for each.

In this respect special mention may be made of the important observation that, in contradistinction to other species, the organism active in the quick acetification process, *Bacterium aceti*, is able to proliferate in a medium containing acetate and ethyl alcohol with ammonium phosphate as the only nitrogen source. Since then this medium, unchanged or only slightly modified, has been used for

---

1) D. P. HOYER, Bijdrage tot de kennis van de azijnbacteriën. Delft, 1898.

2) Initially: *Acetobacterium*.

3) The first instance of this generic name in a printed publication we have been able to trace is to be found in a footnote in the paper on indigo fermentation published in Proc. Kon. Akad. v. Wet. Amsterdam **2**, 495, 1900. A second example occurs in the paper on the lactic acid bacteria in industry published in Arch. néerl. d. sciences exactes et naturelles Sér. II, **6**, 212, 1901.

diagnostic purposes by nearly all investigators who have studied the group under consideration.

In later years BEIJERINCK returned only once to the subject of the acetic acid bacteria. In 1911 he published a paper on pigment formation by acetic acid bacteria in which he described a quite interesting species which unaccountably seems to have escaped the attention of all previous workers in this field [1]). To this species the name of *Acetobacter melanogenum* was given, because it is characterized by its property of producing a dark brown or blackish pigment which resembles melanine in many respects. It is noteworthy that this easily distinguishable species, which in Delft can quite frequently be isolated from beer, does not seem ever to have been encountered by investigators working in other parts of the world.

Although BEIJERINCK's views regarding the nature of the pigment formed probably need revision it seems likely that a further study of *Acetobacter melanogenum* and especially of its pigment production will still lead to interesting results.

1. *On sulphate reduction.*

Soon after the paramount importance of microbial activity for the various chemical conversions proceeding in soil and water had been recognized, the process of nitrate reduction — or denitrification as it is often called — has been the subject of numerous investigations. From various sides valuable contributions to our knowledge of this process have been made. In contrast thereto, the elucidation of the fundamentals of the corresponding process of sulphate reduction has been mainly the work of one man, BEIJERINCK. This statement seems to be especially justified if we include in BEIJERINCK's work the important researches made at the instigation of BEIJERINCK by VAN DELDEN, who was the first to act as an assistant to BEIJERINCK during the latter's academic career.

The origin of hydrogen sulphide in nature had since long attracted attention, and it is not astonishing that sulphates had early been considered as a possible source for it. Between 1864 and 1882, several authors had expressed the opinion that microbes might be agents of the conversion of sulphates into sulphides under natural conditions. However, it was pointed out in 1887 by WINOGRADSKY that the greater part of the organisms which the earlier investigators held responsible for the said conversion were in fact organisms which did not produce hydrogen sulphide, but on the contrary consumed it in their metabolism.

It remained for BEIJERINCK to give in a preliminary paper in 1894 a detailed description of *Spirillum desulfuricans* — nowadays better known as *Vibrio desulfuricans* — the causative organism of sulphate

[1]) Centralbl. f. Bakt. u. Parasitenk. II, **29**, 169, 1911.

reduction [1]). In the two following years more extensive publications appeared in which many different aspects of the problem of biological hydrogen sulphide production were discussed [2]).

BEIJERINCK himself states that the direct inducement to his investigations was of an entirely practical nature. In the yeast factory he was confronted with the problem of freeing the canal water used in the steam boilers from the calcium sulphate present in it. It is typical for BEIJERINCK's originality that he considered in this technical connection the idea of applying a biological method for sulphate destruction. But it is particularly instructive to see that a problem of such restricted scope led to investigations characterized by an exceptional broadness of conception, and dealing exhaustively with the general significance of biological hydrogen sulphide production in nature.

In BEIJERINCK's German publication one reads the following simple sentence: "Die Isolierung des Sulfidfermentes hat mir viel Mühe gekostet". The reasons for his initial failure are then summarised. It is instructive to consider these reasons, because they offer an explanation of the most remarkable fact that even nowadays, 45 years after BEIJERINCK's pioneer work, the number of laboratories in which pure cultures of sulphate-reducing bacteria have been obtained can probably be counted on the fingers of one hand.

It is probably not an exaggeration to state that until very recent years, sulphate reduction had remained practically a special domain entered only by Dutch and Russian investigators.

BEIJERINCK explained that at first he had shared the opinion of the earlier investigators that many of the ordinary aerobic bacteria, occurring in soil and in water, which often display a pronounced reducing activity towards various dyes, would also be able to reduce sulphate. Many experiments, all leading to negative results, had convinced him of the untenability of this view. Careful microscopical examination of well-devised enrichment cultures made him conclude that sulphate reduction proceeded under the influence of a specific organism which, under certain conditions, at least exhibited a typical spirillum-shape. His earlier experiences with species of *Spirillum* led him to the erroneous conclusion that the sulphate reducing spirillum too would be favoured by a certain concentration of free oxygen in the medium. He only gradually realized that the causative organism of sulphate reduction is a strictly anaerobic organism, which in crude cultures, owing to the competition of other bacteria, thrives only in media with low concentrations of simple organic compounds, like lactates, malates, ethyl alcohol, etc.

Yet, even this insight did not remove all difficulties inherent in the

---

1) Versl. Kon. Akad. v. Wet. Amsterdam **3**, 72, 1894.

2) Centralbl. f. Bakt. u. Parasitenk. II, **1**, 1, 49 und 104, 1895. Arch. néerl. d. sciences exactes et naturelles **29**, 233, 1896.

obtaining of pure cultures, as will easily be understood by those bacteriologists who have worked with strictly anaerobic, non-spore-forming bacteria [1]). However, finally BEIJERINCK was successful.

A careful pure-culture study of the exceptional metabolic activities of *Vibrio desulfuricans* was only performed eight years later in collaboration with VAN DELDEN [2]). This investigator was also able to prove that the sulphate reduction which takes place so profusely in brackish water at various spots along the Dutch coast is caused by a bacterium which has again a spirillum or comma shape and which apparently is very closely related to *Vibrio desulfuricans*.

It is noteworthy that from time to time publications appear in which authors claim that organisms which are evidently widely different from *Vibrio desulfuricans* also possess the ability to reduce sulphates. BAARS' monograph on the subject makes it clear that these claims have never been substantiated [3]). In this connection it is also most significant that the study made by ELION [4]) on sulphate reduction under thermophilic conditions led to the conclusion that here too the reduction proceeded under the influence of a vibrio-shaped bacterium, closely related to *Vibrio desulfuricans*. The mass of evidence now available is, therefore, in favour of the view that biological sulphate reduction, the practical importance of which is becoming more manifest every day [5]), is exclusively due to the activity of one of the varieties of a bacterium which was for the first time observed, isolated, and described by BEIJERINCK.

m. *On denitrification.*

As has been observed in the preceding section, BEIJERINCK's contributions to our knowledge of the process of nitrate reduction do not have the same fundamental character as his studies devoted to the process of sulphate reduction. This does not diminish the value of some very remarkable observations made by him upon special features of the denitrification process.

It is greatly to the credit of the French investigators GAYON and DUPETIT to have shown, so early as 1886, that the reduction of nitrates under the influence of a special bacterium led to the formation of nitrous oxide as well as of free nitrogen. This observation had not given rise to any further work till BEIJERINCK took up the question

---

[1]) Only recently STARKEY, working in the Delft microbiological laboratory, has made the startling observation that under certain conditions *Vibrio desulfuricans* is able to form true endospores. Cf. R. L. STARKEY, Archiv f. Mikrobiol. **9,** 268, 1938.

[2]) Arch. d. sciences exactes et naturelles Sér. II, **9,** 131, 1904. A more detailed publication of VAN DELDEN had appeared a year before. Cf. A. H. VAN DELDEN, Centralbl. f. Bakt. u. Parasitenk. II, **11,** 81 und 113, 1903.

[3]) J. K. BAARS, Over sulfaatreductie door bacteriën. Delft, 1930.

[4]) L. ELION, Centralbl. f. Bakt. u. Parasitenk. II, **63,** 58, 1924.

[5]) Cf. C. A. H. VON WOLZOGEN KÜHR and L. S. VAN DER VLUGT, The graphitization of cast iron as an electro-biochemical process in soils. The Hague, 1934.

in 1909. His investigations, made in collaboration with his assistant MINKMAN, were published in the next year [1]).

In the first place a detailed description is given of various enrichment cultures for denitrifying bacteria. This part of the investigation was more or less based on work which VAN ITERSON had performed several years earlier in BEIJERINCK's laboratory [2]). An analysis of the gas developed in these crude fermentations led to the unexpected result that in all cases nitrous oxide was present, although in greatly varying quantities. Especially in experiments with high concentrations of nitrate (8 to 12 per cent) a large percentage of the gas appeared to be nitrous oxide, and BEIJERINCK rigthly emphasized the remarkableness of a biological process leading to the production of a continuous stream of gas containing about 90 per cent of nitrous oxide.

A closer study of denitrification showed that in media of high nitrate concentration two special types of spore-forming bacteria, were active. These unknown denitrifiers could be isolated, and were described under the names of *Bacillus sphaerosporus* and *Bacillus nitroxus*.

The main interest of the paper is, however, to be found in the definite experimental proof that nitrous oxide is not only formed by bacterial activity, but that there are also numerous bacteria which are able to consume this gas. This holds in the first place for many of the denitrifying bacteria themselves, which of course means that nitrous oxide — or the hyponitrous acid from which it is an anhydride — is for these bacteria just an intermediate product in the reduction of nitrate to free nitrogen. But also some bacteria which do not attack nitrates themselves were able to decompose nitrous oxide.

Most striking is finally the demonstration of a new case of "chemosynthesis", namely, the biological production of organic matter from carbon dioxide with the aid of the energy derived from an inorganic reaction. BEIJERINCK showed that a mixture of hydrogen and nitrous oxide makes possible the development of a luxuriant microflora in an inorganic medium containing carbon dioxide. In this case the energy necessary for the carbon dioxide reduction is derived from a reaction between the hydrogen and the nitrous oxide leading to the formation of nitrogen and water. It is clear that this process is quite analogous to the long-known bacterial utilisation of a mixture of hydrogen and oxygen by the so-called hydrogen bacteria.

Another more or less bewildering aspect of denitrification had already been reported by BEIJERINCK in 1903 [3]). A study of the bacteria oxidizing hydrogen sulphide, thiosulphate, etc., as first described by NATANSSOHN, had given BEIJERINCK the conviction that they

[1]) Centralbl. f. Bakt. u. Parasitenk. II, **25,** 30, 1910.

[2]) G. VAN ITERSON Jr., Centralbl. f. Bakt. u. Parasitenk. II, **11,** 689, 1904; Ibid. **12,** 106, 1904.

[3]) Handelingen van het 9e Nederl. Natuur- en Geneeskundig Congres p. 195, 1903; cf. also: Arch. néerl. d. sciences exactes et naturelles Sér. II, **9,** 131, 1904.

were indeed, as claimed by their discoverer, chemo-autotrophic, *i.e.*, that they were able to reduce carbon dioxide with the aid of the energy derived from the oxidation of the sulphur compound. This led BEIJERINCK to the bold idea that there might also be bacteria which could derive the energy necessary for their maintenance and proliferation from an analogous process in which the sulphur compound was oxidized, not with the aid of free oxygen, but with the aid of the oxygen available in nitrates. A further simplification led to the preparation of a fully inorganic medium of which the chief constituents were finely divided sulphur, chalk and nitrate. Herewith an enrichment culture was started in complete absence of free oxygen, and the startling result was obtained that there exist indeed forms of life which can adapt themselves to these extremely primitive conditions.

BEIJERINCK once more returned to this subject in a paper which is the "swan song" of his academic career [1]).

Here many details regarding this remarkable process and its causative organism *Thiobacillus denitrificans* are given. Especially striking is the demonstration that, in this inorganic medium, the formation of organic matter — mostly in the form of bacterial slime — attains such dimensions that it can be demonstrated by the carbonisation reaction which occurs on addition of concentrated sulphuric acid. The paper is concluded by a section in which BEIJERINCK expressed the opinion that *Thiobacillus denitrificans* may well be an autotrophic form of an ordinary heterotrophic denitrifying bacterium like *Bacterium Stutzeri*.

n. *On nitrogen fixation by free-living micro-organisms.*

All students of general and agricultural microbiology are familiar with the association of BEIJERINCK's name with the important subject of nitrogen fixation by free-living micro-organisms. The isolation of *Azotobacter chroococcum* Beijerinck is nowadays a part of the beginner's curriculum in soil microbiology.

A survey of the history of the discovery of this highly remarkable micro-organism is particularly interesting, because it shows clearly that minor factors may largely influence the course of scientific development.

It was BERTHELOT who in 1885 for the first time experimentally proved that the gain in nitrogen which can be nearly always ascertained in fallow land is due to the action of living agents present in the soil. So one can easily understand that, from that time on, several attempts were made to become acquainted with the particular type or types of micro-organisms endowed with the faculty of fixing gaseous nitrogen.

---

[1]) Proc. Kon. Akad. v. Wet. Amsterdam **22,** 899, 1920.

In 1893 this problem attracted the attention of another leader of microbiological thought, WINOGRADSKY, and thus hereby the prospects for a solution might be deemed to be bright.

In the foregoing years WINOGRADSKY had forged a new tool for microbiological work, to wit, the principle of the elective or enrichment culture, and immediately applied this principle with unprecedented success in his researches on the sulphur bacteria and the nitrifying bacteria.

WINOGRADSKY [1]) very naturally decided to proceed in the same way in his efforts to identify the nitrogen fixing bacteria present in soil. He, therefore, prepared culture media free as far as possible from all nitrogen compounds, but containing all other necessary elements, with glucose as a source of carbon and energy, and, moreover, an excess of calcium carbonate. The medium was poured in a thin layer (8–9 mm) in conical flat-bottomed flasks, and after the medium had been inoculated with some soil, a stream of purified air was passed over the cultures. It will be clear that under these conditions luxuriant growth in the medium, especially after a number of transfers to identical media had been made, could only be due to organisms fixing gaseous nitrogen.

In his extensive memoir on the subject which appeared in 1895, WINOGRADSKY indeed succeeded in identifying the organism which predominated in his cultures and found it to be a strictly anaerobic, spore-forming bacterium which provoked a typical butyric acid fermentation [2]). On the ground of its close relation with other butyric acid bacteria the name *Clostridium Pastorianum* was given to the new species. Apparently the development of this anaerobic organism in the enrichment cultures had only been made possible by the simultaneous presence of other bacteria of an aerobic nature in the medium. The pure culture did not develop at all under the conditions of the enrichment culture, that is, in the presence of air. Its nitrogen fixing power was, however, proved beyond doubt, by replacing the air by pure nitrogen. After doing so, a gain in the nitrogen content of the medium could be established with certainty.

WINOGRADSKY was also able to demonstrate the wide distribution of his *Clostridium Pastorianum* in soils of very different origin. By these investigations the question of the nitrogen fixation in arable soils seemed to be solved.

It is impossible to indicate the reasons which made BEIJERINCK five or six years later decide to raise the matter anew. But in a paper [3]) which was first published in 1901 BEIJERINCK opened his introduction with the more or less startling remark:

[1]) Cf. S. WINOGRADSKY, Compt. rend. d. l'Acad. d. Sc. **116,** 1385, 1893; Ibid. **118,** 353, 1894.

[2]) S. WINOGRADSKY, Arch. d. sciences biol. publ. par l'Instit. imp. d. méd. exp. à St. Pétersbourg **3,** 297, 1895.

[3]) Centralbl. f. Bakt. u. Parasitenk. II, **7,** 561, 1901. Later also in: Arch. néerl. d. sciences exactes et naturelles Sér. II, **8,** 190, 1903.

"Unter "Oligonitrophilen" verstehe ich diejenigen Mikroben, welche bei freier Konkurrenz mit der übrigen Mikrobenwelt sich in Nährmedien entwickeln, ohne absichtlich zugefügte Stickstoffverbindungen, aber auch ohne dass Fürsorge getroffen wird, um die letzten Spuren dieser Verbindungen zu entfernen. Sie haben das Vermögen, den freien atmosphärischen Stickstoff binden und zu ihrer Ernährung verwenden zu können."

Herewith, apparently, BEIJERINCK wished to state at once his conviction that nitrogen fixing power is not at all restricted to one or a few specific organisms, but is typical for large groups of microbes.

Characteristic of BEIJERINCK's broad views on the problem is that he also included the photosynthetic organisms in his experiments. In doing so he came to the conclusion, already mentioned in the chapter on the pure culture of the green and the blue-green algae, that the latter group contains several nitrogen fixing species.

The second part of the paper, which deals with the heterotrophic oligonitrophilous organisms, opens with a discussion of WINOGRADSKY's experiments. BEIJERINCK remarks that his own experience led him to the conviction that the development of *Clostridium Pastorianum* is only possible in media which contain small quantities of nitrogen compounds, but this statement does not imply doubt regarding the nitrogen fixing power of the organism, since BEIJERINCK adds that the same holds for the nitrogen fixing organisms discovered by himself.

Then follows a passage which seems sufficiently interesting to be cited again in full:

"Meine Versuche sind von denjenigen von WINOGRADSKY insoweit verschieden, dass ich entweder nur Aërobiose ermöglicht, oder den Sauerstoffzutritt doch in der Weise gefördert habe, dass die Buttersäuregärung unterdrückt, oder sehr geschwächt war. Auch verwendete ich andere Kohlenstoffquellen wie er. Demzufolge kam ich zur Entdeckung einer noch nicht beschriebenen oligonitrophilen Bakteriengattung, welche zu den Aërobien gehört. Ich werde diese durch die Grösse der Individuen leicht kenntliche Gattung *Azotobacter* nennen. Bisher erkannte ich davon 2 sehr verschiedene Arten. Die eine, *A. chroococcum*, ist sehr allgemein in Gartenerde, sowie in allen andern fruchtbaren Bodenarten, die andere ebenso verbreitet im Kanalwasser zu Delft."

It has been deemed interesting to reproduce here in Plate XIII the page of BEIJERINCK's laboratory note-book on which the name *Azotobacter chroococcum* is used for the first time.

By the way it may be remarked that this page gives proof that BEIJERINCK was also in full action on old year's day.

After reading this startling announcement of what since has been proved to be a truly great discovery, one will be eager to learn more details regarding the differences in procedure which decided that

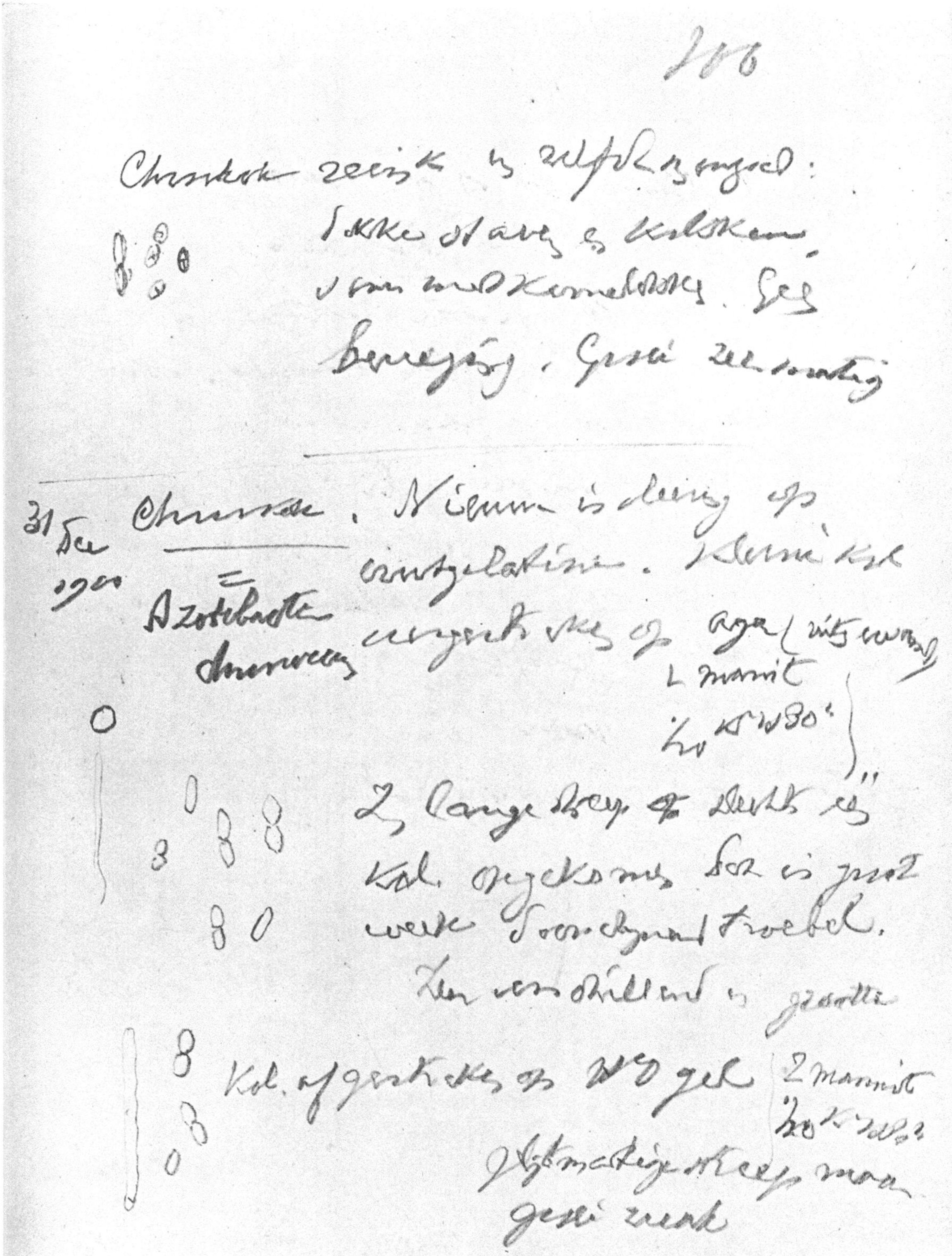

Facsimile of a page of Beijerinck's laboratory note-book (Dec. 31st, 1900).
Here the name *Azotobacter chroococcum* is used for the first time.

experiments made according to exactly the same principle led to so different results in the hands of the two investigators.

BEIJERINCK mentions in the first place as a point of difference that he took measures to promote the aerobic conditions in the medium. However, these measures appear to have been confined to the use of thin layers of culture medium in large Erlenmeyer-flasks, and this was exactly WINOGRADSKY's procedure. It is, moreover, stated explicitly that the mode of renewing the air in the culture flask was the same as in WINOGRADSKY's experiments. So here no explanation of the difference in results can be found.

The second difference in procedure stressed by BEIJERINCK is the use of other carbon sources. BEIJERINCK remarks in this connection that in order to suppress butyric acid fermentation in the medium he has replaced the glucose by substrates, like mannitol and various propionates, the first-named compound being only with difficulty fermentable by butyric acid bacteria, and the propionates not at all.

There seems no doubt that indeed BEIJERINCK's natural tendency to vary widely the composition of the media used by him is directly responsible for his discovery of the new group of nitrogen fixing organisms, which he well may have first observed in media containing one of the substrates mentioned above. However, this explanation is quite inadequate to make comprehensible why WINOGRADSKY should not have observed the same organisms six years earlier. For although BEIJERINCK rightly maintains that media containing mannitol or propionate have the advantage that in these media the anaerobic spore-forming organisms develop more slowly than in glucose media, yet, every student of soil microbiology will be prepared to confirm that as a rule *Azotobacter* develops in an equally abundant way in enrichment cultures made with media containing glucose and calcium carbonate.

This point of view is fully confirmed by WINOGRADSKY himself. BEIJERINCK's communication seems to have revived his interest in the problem in question, for the next year he published in the "Centralblatt für Bakteriologie" another extensive memoir on *Clostridium Pastorianum* [1]).

As motive for this sudden activity after seven years of silence WINOGRADSKY mentions that he often received inquiries from colleagues regarding the identity of certain strains with *Clostridium Pastorianum* and thus concluded that the description of the said species in his 1895 paper was not sufficiently complete. He then gives a very detailed survey of the morphological and fermentation properties of the organism. In connection with the question under discussion the supplement is by far the most interesting part of the publication. Herein he gives his reflections on BEIJERINCK's recent publication. In

---

1) S. WINOGRADSKY, Centralbl. f. Bakt. u. Parasitenk. II, **9**, 43 und 107, 1902.

the first place he rejects BEIJERINCK's designation "oligonitrophilous" as far as *Clostridium Pastorianum* is concerned, sufficient proof having been given that this organism is able to proliferate indefinitely in the complete absence of nitrogen compounds. The passages dealing with BEIJERINCK's *Azotobacter* discovery are at the same time so characteristic and so instructive that it seems justified to cite them here in full:

"Die *kleine* aërobe Bakterienflora, welche in zuckerhaltigen, stickstoffarmen Nährlösungen auftritt, *ist mir seit 1893 bekannt.* Sie entwickelt sich *ganz konstant als Kahmhaut* auf den abgegorenen Kulturen, tritt manchmal aber auch selbständig auf in den für *Clostridium Pastorianum* bestimmten, aber nicht gärenden Kulturen. Es gelang meistens ohne Mühe durch einfache mechanische Mittel, diese Arten von *Clostridium Pastorianum* zu trennen und aus einer Mutterkultur zwei Reihen — eine gärende und eine nicht gärende — herauszuzüchten. Abgesehen von 2 oder 3 sporenbildenden Bazillen, finde ich *in meinen Tagebüchern beschrieben und abgebildet 1. einen "Sarcina-ähnlichen" oder "Chroococcus-ähnlichen" Organismus (beide Bezeichnungen werden gebraucht), als häufigste Erscheinung, welcher anfangs eine weissliche, etwas irisierende, schliesslich braun werdende Membran bildet,* 2. ein kurzes dickliches Spirillum...."

and somewhat further on:

"Alle diese Formen zogen meine Aufmerksamkeit auf sich *in allen Böden, die ich untersuchte,* sowohl in Petersburger und den südrussischen, wie auch im Pariser. Dieses ihr konstantes Auftreten unter Bedingungen, in welchen scheinbar nur gasförmigen Stickstoff assimilierende Arten gut gedeihen könnten, *erweckte oft meinen Verdacht, ob sie sich nicht an dem Vorgange der Stickstoffassimilation beteiligen könnten.* Da aber andererseits ihr Wachstum im Vergleiche mit *Clostridium Pastorianum* doch wenig imponierend erschien, da ich weiter schon eine Anzahl von Mikrobien kannte, die unter diesen Bedingungen zwar Wachstumserscheinungen, aber zweifelhafte Assimilationsfähigkeit dem atmosphärischen Stickstoff gegenüber zeigten, *so habe ich ihnen kein weiteres Interesse geschenkt* und keine Musse gefunden, sie näher zu untersuchen" [1]).

These citations do not leave doubt that WINOGRADSKY had forestalled BEIJERINCK in his *Azotobacter* discovery by at least 7 or 8 years. But at the same time the further development of soil microbiology has definitely proved that WINOGRADSKY had grievously failed to appreciate the great significance of an organism which apparently had been so abundant in his enrichment cultures.

One may ask why BEIJERINCK reacted so differently to the regular appearance of *Azotobacter* in his cultures. Just like WINOGRADSKY he was impressed with the inadmissibility of ascribing to every organism growing in a so-called nitrogen-free medium the faculty of nitrogen

---

1) Italics in these citations are mine (A. J. K.).

fixation. One might thus expect that he would not have expressed his firm belief in the great importance of *Azotobacter chroococcum* without having convinced himself that the cultivation of this species indeed leads to a noticeable gain in nitrogen of the medium. However, in BEIJERINCK's first publication there is no indication that he even attempted to do so. In this connection WINOGRADSKY rightly remarked:

"Obgleich wir nun, Dank BEIJERINCK, die genaue Charakteristik dieser Arten jetzt besitzen, bleibt doch immer der wichtigste Punkt noch unaufgeklärt, nämlich ob dieselben atmosphärischen Stickstoff assimilieren können oder nicht. Die blosse Thatsache ihres Vorkommens in stickstoffarmen Nährlösungen beweist natürlich nichts."

In view of all this there remains only one explanation for BEIJERINCK's discovery, namely, intuition or even better, genius! And if WINOGRADSKY in 1893 failed to deal adequately with the situation, the reason can only be that at that time his genius had been too much captivated by his great discovery of *Clostridium Pastorianum*.

After this circumstantial historical introduction to the *Azotobacter* discovery only a few more remarks will be made on BEIJERINCK's further contributions to the problem of microbial nitrogen fixation. In the first place it should be emphasized that the way in which he described the various stages of development of *Azotobacter chroococcum* is exemplary. It is noteworthy that he succeeded in completely avoiding the pit-falls of which several later investigators have been become the victims.

Then it is characteristic for BEIJERINCK's universality and thoroughness that already in his first publication he described a second, clearly distinct species of his new genus, *viz.*, *Azotobacter agilis*. BEIJERINCK found that this second species, with its much larger cells, usually predominates in the enrichment cultures, if canal water, instead of soil, is used for the inoculation. In a fairly recent paper, published 32 years after the discovery of *Azotobacter agilis*, it was concluded that this organism had until that time not been isolated except from Dutch canal waters [1]). It seems probable that this second *Azotobacter* species which also exhibits a good nitrogen fixing power, is of material significance for the economy of fresh-water communities, at least, in those regions in which the water is not free from pollution.

Of the later publications of BEIJERINCK on "oligonitrophilous microbes" we pass over those dealing with the photo-synthetically active microbes, because they have been considered in Chapter XVI.

---

[1]) A. J. KLUYVER und W. J. VAN REENEN, Archiv f. Mikrobiol. **4**, 280, 1933; cf. also A. J. KLUYVER und M. T. VAN DEN BOUT, Ibid. **7**, 261, 1936.

It is interesting to add that since the appearance of the first paper, HUGH NICOL, at Rothamsted, isolated a strain of *A. agilis* from a drainage ditch at Oby Mill, Norfolk, England (Private communication; cf. E. J. RUSSELL, Soil Conditions and Plant Growth, 7th Ed., 1937, p. 384). More recently WINOGRADSKY has also isolated typical strains of *A. agilis* from surface waters in France. Cf. S. WINOGRADSKY, Ann. de l' Inst. Pasteur **60**, 351, 1938.

The extensive paper on nitrogen fixation which BEIJERINCK and his collaborator VAN DELDEN published in 1902 [1]) need not to be reviewed here in detail. The paper sets out extensive data regarding the gain of nitrogen in cultures in which *Azotobacter* was growing together with other "oligonitrophilous" species. The conclusion was that *Azotobacter* itself is unable to fix nitrogen and that its proliferation in the enrichment cultures is exclusively due to its living in symbiosis with other nitrogen fixing species. This view has now been definitely refuted by the work of numerous other investigators. Nevertheless the paper remains of interest on account of its detailed description of the many other bacteria which regularly accompany *Azotobacter* in the enrichment cultures.

Six years later BEIJERINCK returned once more to the subject [2]). In this publication he revoked his opinion regarding the absence of nitrogen fixing power in *Azotobacter*. This time, in collaboration with his assistant MINKMAN, definite proof for nitrogen fixation in pure cultures was given. In a final section of the paper a few observations are recorded regarding the distribution of *Azotobacter* in soil. The procedure applied, *viz.*, the direct sowing of soil particles on elective solid media, has later in the hands of WINOGRADSKY proved to be a most valuable tool in soil microbiology [3]).

Finally, mention may here be made of a short paper — published only in the Dutch language — which BEIJERINCK wrote in the last year of his academic career [4]). Herein he gave his views on the significance to be attached to the more or less frequent occurrence of *Azotobacter* in soils. BEIJERINCK seemed inclined to conclude that the number of *Azotobacter* cells detectable in soil would be an indicator of its fertility. In contrast hereto he placed the observation that *Granulobacter Pastorianum* is equally frequent in fertile and infertile soils. Although the data on which these conclusions are based are too scanty to lend them more than a provisional character, the paper has the merit of inciting further research in this direction.

o. *Investigations on urea-decomposing bacteria.*

As has already been observed in Chapter XX it was at the beginning of this century that BEIJERINCK became fully aware of the far-reaching importance of the principle of the enrichment culture. His study on the group of the urea-decomposing bacteria which appeared in 1901 and in which he for the first time made more general remarks on the said principle, also afforded a splendid demonstration of what can be attained by a well-designed application thereof [5]).

1) Centralbl. f. Bakt. u. Parasitenk. II, **9,** 3, 1902.
2) Proc. Kon. Akad. v. Wet. Amsterdam **11,** 67, 1908.
3) Cf. Ann. de l'Inst. Pasteur **40,** 455, 1926.
4) Versl. Kon. Akad. v. Wet. Amsterdam **30,** 431, 1921.
5) Centralbl. f. Bakt. u. Parasitenk. II, **7,** 33, 1901.

BEIJERINCK's sterling merit appears from a comparison of his results with those of earlier investigators in the field of urea decomposition, like VAN TIEGHEM, MIQUEL, VON JACKSCH, and LEUBE. It is true that especially MIQUEL had added a good deal to our knowledge of the process in question, nevertheless the greater part of his observations bear an incidental character. On the contrary the prescriptions given by BEIJERINCK for the accumulation of various urea bacteria lead in many cases to reproducible results, thus offering a firm foundation for our knowledge of this group of bacteria.

It is of no use to enter here into details regarding the various accumulation experiments described. In the hands of BEIJERINCK they led to the isolation of the following species: *Urococcus ureae* Cohn *Urobacillus pasteurii* Miquel, *Urobacillus miquelii* nov. spec., *Urobacillus leubei* nov. spec., and *Planosarcina ureae* nov. spec. Careful descriptions were given of all these species, ably supported by beautiful drawings. Special attention was given to the degree to which these species differ in urea-decomposing activity; *Urobacillus pasteurii*, which is able to decompose not less than 10 per cent urea present in its medium, bears the palm in this respect. It should be realized that this means vital activity in a medium containing finally about 13 per cent of ammonium carbonate! Probably this is the upper limit for alkali concentration tolerated by a living organism.

Another culmination point in the publication is the discovery of *Planosarcina ureae*, a gem of the microbe world. It is well known that motile cocci are very rare, and the finding of a motile coccus-shaped bacterium forming regular tetrads must, therefore, be deemed a first rate discovery. But the further circumstance that this organism presented the first indubitable case of formation of endospores in a non-rod-shaped bacterium meant nothing short of a revolution in the current views on bacterial morphology and life cycles.

The exceptional character of *Planosarcina ureae* was apparently heightened by a circumstance of a secondary nature. Several of BEIJERINCK's pupils, in later years, found that the accumulation experiment as prescribed by BEIJERINCK for *Planosarcina ureae* always gave negative results. At one time attempts at its isolation were made simultaneously in Delft, Amsterdam, Haarlem and Wageningen, but in all cases the *Planosarcina* failed to appear. This has led to the suspicion that the bacterium in question with its strongly abnormal morphology might have to be considered as a disappearing species the last representatives of which had incidentally been encountered by BEIJERINCK.

A few years ago, however, this view was shown to be untenable by GIBSON who demonstrated the ubiquity of *Sarcina ureae* in soil. GIBSON used a procedure based on principles quite different from the original method described by BEIJERINCK [1]). By applying GIBSON's

[1]) T. GIBSON, Archiv f. Mikrobiol. **6**, 73, 1935.

method, the presence of *Planosarcina ureae* in various Dutch soils could easily be demonstrated.

The discussion of BEIJERINCK's memoir on the urea bacteria would be incomplete, if no reference was made here to the elegant and simple procedure which he devised as a quick test of urea-decomposing ability applicable to various microbes, or to vegetable and animal tissues. It is sufficient to place some of the material to be tested on the surface of a gelatine plate which contains 12 per cent of gelatine, yeast extract, and 2 or 3 per cent urea. If the test material converts the urea into ammonium carbonate, one observes after a few minutes in the surface of the gelatine directly surrounding the test material a very fine precipitate, formed initially in a very thin layer. On looking at the plate at a certain angle the precipitate manifests itself clearly by the formation of beautiful Newton diffraction rings, BEIJERINCK has given the name of "iris-phenomenon" to the effect. It is easily shown that the phenomenon is primarily due to the formation of ammonium carbonate by the bacteria, the direct application of the said salt giving at once the same effect. The precipitate ultimately formed is probably a mixture of calcium carbonate and calcium phosphate [1]; for some reason or other the precipitation begins at the surface of the gelatine gel.

Thanks to this very sensitive, yet simple reaction, BEIJERINCK was in later years able to demonstrate the presence of urea-decomposing power in several strains of root nodule bacteria [2]. The significance of this finding has not yet been elucidated.

p. *Bacillus oligocarbophilus, an agent of the biological purification of the air.*

At some time BEIJERINCK observed the development of a quite specific microflora in a medium which only contained small quantities of nitrate, phosphate and traces of salts of magnesium, manganese and iron. This surprizing phenomenon led to a careful study, made in collaboration with his assistant VAN DELDEN, the results of which were published in 1903 [3]. Since the experiment had been performed in the dark, and, therefore, light was not an energy source, the problem arose at once from where the energy necessary for the development of this flora originated. If such a source could be indicated it would be, of course, possible to ascribe the origin of the organic material, accumulating in this inorganic medium, to a reduction of the carbon dioxide of the air. It is well known that the nitrifying bacteria, for example, are able to convert carbon dioxide into cell material with the

---

[1] The yeast extract always contains a small amount of soluble calcium salts.
[2] Nature **112,** 439, 1923.
[3] M. W. BEIJERINCK und A. VAN DELDEN, Centralbl. f. Bakt. u. Parasitenk. II, **10,** 33, 1903.

aid of the energy derived from the oxidation of ammonia or of nitrite.

But in BEIJERINCK's particular experiment, the nitrogen had been added to the medium in its highest stage of oxidation — as nitrate — and for this reason at first sight no energy source could be traced. Nevertheless there remained the undeniable fact that the media in question after inoculation with some soil were fairly soon covered with a thin, white or feebly rose-coloured very dry film consisting of minute bacteria stuck together by a slimy substance. This organism, to which the name of *Bacillus oligocarbophilus* was given, could without any difficulty be transferred into fresh culture media, and the cultures so obtained could be kept going indefinitely. By chemical analysis it was shown convincingly that in such cultures very considerable amounts of carbon accumulated in the media, and since these could not be derived from the carbon dioxide of the air, the conclusion was inevitable that unknown organic compounds present in the polluted air of the laboratory — and in general in the air of all inhabited dwellings — were directly responsible for the proliferation of *Bacillus oligocarbophilus*. In agreement herewith it was shown that practically no development took place in the much purer air of a greenhouse. Apart from acting as carbon food the said impurities must also serve as a substrate for the respiration of the bacterium and thus partly be converted into carbon dioxide. It is clear that all this means that the organism in question acts as a powerful agent of air purification, a process which forms an interesting counterpart to the well-known processes of water purification.

The interest of these findings is manifold. In the first place, the mode of discovery of *Bacillus oligocarbophilus* is a very fine example of what may be called "a perfect accumulation experiment", *i.e.*, a case in which enrichment experiments in the highly elective medium led after a very few transfers to an almost pure culture [1]). Secondly, it shows that it is possible to demonstrate in our everyday atmosphere the presence of not-negligible amounts of organic substances which are usually overlooked. This implied that the surrounding air is a potential source of microbial life which may manifest itself where it has not been expected. This may lead to erroneous conclusions with regard to the nature of a microflora present under special conditions. It is easily understood that if one finds an abundant development of a certain microbe in a fully inorganic medium containing nitrite one will be inclined to consider this compound as the energy source of the vegetation. It is nevertheless possible that the development is due to the organic energy sources present in impure air. It seems probable that even in recent studies on nitratation this point of view has been

---

[1]) Some reserve seems indicated here, since KINGMA BOLTJES recently found in *Hyphomicrobium vulgare* a second agent of air purification with closely related physiological properties. Cf. T. Y. KINGMA BOLTJES, Archiv f. Mikrobiol. **7**, 188, 1936.

lost sight of, and a perusal of BEIJERINCK and VAN DELDEN's study can be recommended to any microbiologist.

It may finally be remarked that the question of the systematic relationships of *Bacillus oligocarbophilus*, on which point some very fallacious views have been ventilated in the literature, is greatly in need of reconsideration.

q. *Studies on microbial variation* [1]).

Such a keen observer as BEIJERINCK was could not have failed to be struck — even very early in his career — by the phenomena of variation occurring with the various microbes which he studied in detail. As might therefore be expected, the places in BEIJERINCK's papers in which he refers to such variations are numerous. This review will, however, be restricted to those publications in which BEIJERINCK makes an attempt to collect and to co-ordinate his various experiences in this field.

We may start with the lecture which BEIJERINCK held in the meeting of the Koninklijke Akademie van Wetenschappen of Amsterdam on October 27th, 1900 [2]). The lecture was, as stated by BEIJERINCK himself, a direct consequence of the fact that a month before HUGO DE VRIES at the same place had dealt with the origin of new forms in higher plants in a lecture which brought a first outline of his well-known mutation theory.

In the introduction BEIJERINCK expounds the advantages which micro-organisms offer for the investigation of the laws of heredity and variability, but it has to be acknowledged that nowadays it is difficult to subscribe to several of his arguments.

On proceeding to the subject proper — the different forms of hereditary variation of microbes — BEIJERINCK makes a plea for his view that mainly three types of variation should be distinguished, *i.e.*, degeneration, transformation and "common" variation.

The term "degeneration" applies to the case that a freshly isolated culture — initially growing abundantly — gradually and successively loses, various properties this process finally leading to a complete loss of reproductive power. The bacterium of "long whey", *Streptococcus hollandiae*, which on cultivation rapidly loses its ability of slime production, and which on prolonged cultivation quite regularly dies off, is offered as an example.

The word "transformation" is used in those cases in which all individual cells present in a culture undergo a common change — usually a loss — in properties. The loss of luminescence regularly oc-

[1]) The reader is also referred to the interesting survey of J. J. VAN LOGHEM, Beijerinck en de kennis der bacterieele veranderlijkheid (Ned. Tijdschr. v. Geneesk. **75,** 1046, 1931).

[2]) Proc. Kon. Akad. v. Wet. Amsterdam **3,** 352, 1900; Arch. néerl. d. sciences exactes et naturelles Sér. II, **4,** 213, 1901.

curring in a culture of *Photobacterium luminosum* is given as one of the examples.

Finally the term "variation" is reserved for those cases in which the original form is maintained, whilst, now and then, individual cells are thrown off with different properties which on the whole are likewise constant and remain so. Only occasionally the new forms throw off other variants, amongst which the normal form may occur as an atavist. A detailed description of several examples of this variation in the more restricted sense is given in the paper.

The discussion which followed BEIJERINCK's lecture, in which discussion DE VRIES also took part, apparently induced BEIJERINCK to add to his paper a foot-note in which he says to agree perfectly with the opinion of DE VRIES that sudden variation — mutation — is often responsible for the origin of new species. However, he emphasizes that this concept is not capable of explaining the adaptation which so often is characteristic for the variation.

In 1911, in the first meeting of the "Nederlandsche Vereeniging voor Microbiologie", BEIJERINCK returned to the subject. The extensive paper which was published as a result of this, his presidential address, is before all remarkable for its wealth of observations on the variation of several very dissimilar micro-organisms [1]).

Even to-day any theory of microbial variation should take account of the numerous experimental data collected by BEIJERINCK.

For BEIJERINCK himself these observations formed an ample basis for his theoretical considerations, which deviate in many respects from his earlier ones.

This time BEIJERINCK distinguished three types of microbial variation, *viz.*, modification, fluctuation and mutation.

"Modification" is the variation which may occur, if a microbe is brought under a certain set of external conditions, but which disappears, either at once or after a few cell generations, as soon as the primary conditions are restored. This form of variation is, therefore, non-hereditary. "Fluctuation" is the term used for the hereditary change which may take place under the influence of external conditions, in so far as this change is characterized by the fact that all or the great majority of the individual cells of a strain are subject to it. In "mutation" the external conditions are of subordinate importance, the principal factors are the internal conditions present in a relatively small number of cells.

However, since fluctuations also occur leap-wise and external conditions are sometimes decisive for mutations as well, there is only a difference in degree between the two latter types of variation.

The main part of the paper is devoted to a minute description of the variation phenomena observed with various microbial cultures. It is

---

1) Folia Microbiologica **1**, 1, 1912.

characteristic of BEIJERINCK's versatility that amongst these cultures there are three bacterial species, *viz.*, *Bacillus prodigiosus*, *Bacillus herbicola* and *Bacillus indicus*, one alga: *Chlorella variegata*, and a few yeasts amongst which *Schizosacharomyces octosporus* is especially considered.

In the final chapter of his paper BEIJERINCK deals exhaustively with the nature of the variations observed. He concludes that the majority of these variations must be considered as mutations which are wholly comparable to the more or less constant bud mutations of higher plants. He also draws a parallel between microbial mutations on the one hand and the occurrence of different forms of heterostyles, and that of the two sexes of dioecious plants on the other hand. But also the formation of the different organs in higher organisms — a phenonemom usually simply designated as differentiation — is considered to present a more or less analogous case [1]).

In identifying microbial variations with the well-known gene mutations of higher organisms BEIJERINCK, of course, is well aware of the fact that in micro-organisms no experimental proof for the correctness of this assumption can be furnished, owing to the impossibility of a gene analysis by hybridization. Yet he emphasizes that there is no reason to accept that mutants of organisms showing amphimixis should in any respect be different from those with asexual reproduction only.

A characteristic feature of BEIJERINCK's views is his conviction that mutation and atavismus are equivalent processes.

According to BEIJERINCK many mutation phenomena should be regarded as to be of an atavistic nature. This may even apply, when apparently a new property as, for instance, pigment production is manifested. This may merely mean that a progene is brought back into the active state. In other cases active genes may be reverted into progenes.

It is here not the place to enter into a detailed consideration of the fate of the mutation theory of microbial variation during the quarter of a century that has passed, since BEIJERINCK gave his masterly exposé. Suffice it to state that many of the later investigators in this field have severely criticized BEIJERINCK's views. Other theories, amongst which VAN LOGHEM's "individuality theory" [2]) and HADLEY's cyclic theory [3]) may be especially mentioned, have largely superseded the mutation concept. Of late, however, both LINDEGREN [4]) and

---

[1]) In a recent survey of the variability of bacteria this point of view has again been brought to the fore by O. RAHN. Cf. Scientia, 1937, p. 83.

[2]) J. J. VAN LOGHEM, Nederl. Tijdschr. v. Geneesk. **65,** 2981, 1921; Proc. Kon. Akad.v. Wet. Amsterdam **34,** 2, 1931; Antonie van Leeuwenhoek **4,** 113, 1937.

[3]) PH. HADLEY, Journ. of Infect. Dis. **40,** 1, 1927; Ibid. **48,** 1, 1931; Ibid. **60,** 129, 1937.

[4]) C. C. LINDEGREN, Zentr. f. Bakt. II, **92,** 40, 1935; Ibid. **93,** 113, 1936.

DESKOWITZ [1]) have again forwarded important arguments in favour of the view that microbial variation is indeed largely due to gene mutation, and the same holds also for MAYER [2]), to whose up to date survey of the problem in question the reader may be referred.

There is, however, still another contribution of BEIJERINCK to our knowledge of the variation problem which may not pass unmentioned. In 1914 BEIJERINCK published a paper which bore the title: "On the nitrate ferment and on the physiological formation of species" [3]). He reported in this paper his experiences, undoubtedly collected over numerous years, regarding the nitrate ferment. On the whole his observations are in substantial agreement with the results of WINOGRADSKY'S classical study which appeared as long back as 1890. However, BEIJERINCK added one new feature to the picture drawn by the Russian scientist. He gave it as his conviction that, contrary to the current opinion, the nitrate ferment was quite capable of proliferation in common media rich in organic substances. But on doing so, its ability to oxidize nitrites was irreparably lost. Out of the oligotrophic nitrate ferment, *Nitrobacter oligotrophum*, a new species, *Nitrobacter polytrophum*, was irreversibily formed, hence the term "physiological formation of species". It will be clear that it is extremely difficult to arrive at a final decision regarding the correctness of this theory. For the irreversibility of the assumed conversion makes it almost impossible to disprove that the so-called polytrophic form is not actually a common heterotrophic contaminant which has maintained itself in the cultures of the nitrate ferment during its cultivation in the inorganic media.

It is, therefore, not surprizing that WINOGRADSKY has severely criticized BEIJERINCK'S observations and in consequence has fully rejected his theory of physiological species formation [4]). It may be added that the results of the recent investigations of KINGMA BOLTJES are also against BEIJERINCK'S ideas [5]).

Yet, it seems wise not to lose sight of the fact that the more or less startling observations in question were made by a BEIJERINCK in the last phase of his career, that is to say by a microbiologist who was not likely to be led astray by common contaminants. Moreover, again according to BEIJERINCK, this example of physiological species formation did not stand alone. In the last paper which he published before his retirement from the chair, BEIJERINCK described a similar phenomenon for the bacterium active in the process of denitrification with sulphur as a source of energy [6]). On transference into organic

---

[1]) M. W. DESKOWITZ, Journ. of Bact. **33,** 349, 1937.

[2]) H. D. MAYER, Das Tibi-Konsortium, nebst einem Beitrag zur Kenntnis der Bakterien-Dissoziation. Delft, 1938.

[3]) Folia Microbiologica **3,** 91, 1914.

[4]) Compt. rend. de l'Acad. d. Sc. **175,** 301, 1922.

[5]) T. Y. KINGMA BOLTJES, Archiv f. Mikrobiol. **6,** 79, 1935.

[6]) Proc. Kon. Akad. v. Wet. Amsterdam **22,** 899, 1920.

media, this organism should become irreversibly converted into the common denitrifying species, *Bacterium Stutzeri.*

It seems probable that younger microbiologists will be inclined to cover these later publications of BEIJERINCK with the cloak of charity; older workers in the field who are more familiar with the tricky ways in which variation may manifest itself will be led to wonder: senescence or accumulated wisdom?

THE ENVOY

In concluding this survey of BEIJERINCK's main contributions to the science of microbiology the author is fully aware of the incompleteness of the picture drawn up.

Yet he ventures to hope that the light thrown upon the versatility, the originality, and the vastness of BEIJERINCK's studies in the microbiological field will have been sufficiently strong to establish the conviction that such a work could only be performed by a man whose life has been completely devoted to the pursuit of knowledge, and to the search for scientific truth.

If the author has succeeded in this, he will have achieved a task which has been badly neglected by BEIJERINCK himself. DE KRUIF writes in his "Microbe Hunters": "There have been searchers who have failed — they have kept on hunting with the naturalness of ducks swimming; there have been searchers who have suceeded gloriously — but they were hunters born, and they kept on hunting in spite of the seductions of glory." It will be difficult to find any one for whom the last part of this dictum holds better than for BEIJERINCK.

Unaffected by the numerous honours bestowed upon him, BEIJERINCK offers the picture of a man whose life was entirely ruled by a craving for knowledge. Neither fatigue nor compromise existed for him: his never-saturated mind drove him from one problem to another, and a life resulted so fully devoted to science that no space for celebrity was left therein.

BEIJERINCK always abandoned to others the task of disseminating his knowledge; he sought only — to speak once more with the words of DE KRUIF — "that priceless loneliness that is the one condition for all true searching."

Perhaps BEIJERINCK's attitude of mind cannot be better characterized than by stating that, when he addressed the students at the occasion of the opening of his laboratory on September 28th, 1897, he chose to conclude with the following quotation from PASTEUR:

"Vivez dans la paix séreine des laboratoires et des bibliothèques. Dites vous d'abord: "Qu'ai-je fait pour mon instruction?" "Puis à mesure que vous avancerez "Qu'ai-je fait pour mon pays?" jusqu'au moment où vous aurez peut-être cet immense bonheur de penser que vous avez contribué en quelque chose au progrès et au bien de l'humanité. Mais, que les efforts soient plus ou moins favorisés par la vie, il faut, quand on approche du grand but, être en droit de se dire: J'ai fait ce que j'ai pu"."

Verily, these last words would be the fitting epitaph for BEIJERINCK.

APPENDICES

Appendix A.

The "Stellingen" accompanying BEIJERINCK's doctorate thesis. *)

STELLINGEN.

I.

De stof is vortex-vibratie van den aether (WILLIAM THOMSON).

II.

Voor de verdere ontwikkeling der spectraal-analyse is het wenschelijk dat men nauwkeuriger bekend worde met den graad van dissociatie van verschillende lichamen bij verschillende temperaturen.

III.

Door de onderzoekingen van VICTOR MEIJER is de vijfwaardigheid van de stikstof niet bewezen.

IV.

Ten onrechte beweert FITTIG dat de isomerie van fumaar- en maleïnzuur beter verklaard kan worden door het aannemen van twee vrije affiniteiten van de koolstof dan door VAN 'T HOFF's hypothese.

V.

Protoplasma uit somtijds werking op afstand.

VI.

De onderzoekingen van ADOLF MAYER leveren het bewijs, dat zekere Crassulaceën zuurstof kunnen afscheiden ook buiten de aanwezigheid van koolzuur.

VII.

Niet altijd is levend protoplasma ondoordringbaar voor kleurstoffen.

*) Some obvious printing errors occurring in the original text have been corrected.

VIII.

De oudste organismen waren bladgroenhoudend.

IX.

Een langdurig voortgezette vermenigvuldiging van Phanerogamen zonder geslachtelijke voortplanting kan tot uitsterving leiden.

X.

Onjuist is DARWIN's beweren (Domestication II p. 255): „if it were possible to expose all the individuals of a species during many generations to absolutely uniform conditions of life, there would be no variability."

XI.

Saccharomyces is een Ascomyceet.

XII.

De door MÜLLER (Thurgau) „Blattvertreter", genoemde aanhangselen van het protonema der bladmossen hebben niet de waarde van phyllomen.

XIII.

De richting van den eersten deelwand in de eicel der archegoniaten is voor hun rangschikking van geen hooge waarde.

XIV.

Asterophyllites kan met meer recht tot de Lycopodiaceën dan tot de Calamariën worden gerekend.

XV.

De Monocotylen zijn nader verwant aan Isoëtes dan aan de Dicotylen.

XVI.

Phanerogamen kunnen twee of meer vaders gelijktijdig bezitten.

XVII.

De gelede meeldraad van Euphorbia is geen enkelvoudige meeldraad.

XVIII.

De postembryonale ontwikkeling der insekten is geen weerspiegeling van hun phylogenie.

XIX.

Siredon stamt af van Amblystoma.

XX.

De toestand van entropie van het heelal is onbereikbaar.

Appendix B.

List of BEIJERINCK's assistants in his academic period.

A. H. VAN DELDEN	1 September 1895	—1 September 1904
G. VAN ITERSON JR.	4 September 1902	—1 September 1907
H. C. JACOBSEN	1 September 1904	—1 Maart 1916
D. C. J. MINKMAN	1 September 1907	—1 September 1911
N. L. SÖHNGEN	1 December 1911	—1 September 1915
T. FOLPMERS	1 Januari 1916	—1 Januari 1917
Mej. J. E. VAN AMSTEL	1 Juni 1916	—1 September 1916
J. DE GRAAFF	6 December 1916	—1 November 1919
W. BEIJERINCK	16 Januari 1917	—1 September 1918
Mej. J. C. MEISS	3 December 1918	—1 Februari 1920
J. VAN BEYNUM	1 Januari 1920	—1 December 1920
L. E. DEN DOOREN DE JONG	1 Mei 1920	—(16 Augustus 1923)
H. J. L. DONKER	1 Juni 1921	—(1 Juni 1924)

Appendix C.

List of communications from the laboratory for microbiology at Delft, published by BEIJERINCK's collaborators in the years 1895—1921. [1])

A. H. VAN DELDEN

Ein Hülfsapparat zur Einstellung mit Immersions-objectiven. Z.f. wissensch. Mikrosk. u. f. mikrosk. Techn. **12,** 15 (1895).

Beitrag zur Kenntnis der Sulfatreduktion durch Bakterien. Centralbl. f. Bakt. II, **11,** 81 und 113 (1904).

H. TER MEULEN

De bepaling van mosterdolie in raapkoeken. Handel. van het 8ste Nederl. Natuur- en Geneesk. Congres, 88 (1901).

H. H. GRAN

Studien über Meeresbakterien. I. Reduction von Nitraten und Nitriten. Bergens Museums Aarbog 1901, No. 10, p. 1.

Studien über Meeresbakterien. II. Ueber die Hydrolyse des Agar-Agars durch ein neues Enzym, die Gelase. Bergens Museums Aarbog 1902, No. 2, p. 1.

C. J. J. VAN HALL

Bacillus subtilis (Ehrenberg) Cohn und *Bacillus vulgatus* (Flügge) Mig. als Pflanzenparasiten. Centralbl. f. Bakt. II, **9,** 642 (1902).

G. VAN ITERSON JR.

L'acide carbonique atmosphérique. Ann. de l'Obs. municipal de Montsouris **3,** 372 (1902).

Ophoopingsproeven met denitrificeerende bacteriën. Versl. Kon. Akad. v. Wetensch. A'dam **11,** 135 (1902).

Accumulation experiments with denitrifying bacteria. Proc. Kon. Akad. v. Wetensch. A'dam **5,** 148 (1902).

De aantasting van cellulose door aerobe mikro-organismen. Versl. Kon. Akad. v. Wetensch. A'dam **11,** 807 (1903).

The decomposition of cellulose by aerobic micro-organisms. Proc. Kon. Akad. v. Wetensch. A'dam **5,** 685 (1903).

---

[1]) Cf. also Appendix D.

Die Zersetzung von Cellulose durch aerobe Mikro-organismen. Centralbl. f. Bakt. II, **11,** 689 (1904).

Over denitrificatie. Chem. Weekbl. **1,** 691 (1904).

Anhäufungsversuche mit denitrifizierenden Bakterien. Centralbl. f. Bakt. II, **12,** 106 (1904).

Over den kringloop der zwavel in de organische natuur. 14e Jaarverslag Technol. Gezelschap, 57 (1905).

N. L. SÖHNGEN

Over bacteriën, welke methaan als koolstofvoedsel en energiebron gebruiken. Versl. Kon. Akad. v. Wetensch. A'dam **14,** 289 (1905).

Methan as carbon-food and source of energy for bacteria. Proc. Kon. Akad. v. Wetensch. A'dam **8,** 327 (1905).

Ueber Bakterien, welche Methan als Kohlenstoffnahrung und Energiequelle gebrauchen. Centralbl. f. Bakt. II, **15,** 513 (1906).

Ureumsplitsing bij afwezigheid van eiwitten. Versl. Kon. Akad. v. Wetensch. A'dam **17,** 348 (1908).

The splitting up of ureum in the absence of albumen. Proc. Kon. Akad. v. Wetensch. A'dam **11,** 513 (1909).

Ureumspaltung bei Nichtvorhandensein von Eiweiss. Centralbl. f. Bakt. II, **23,** 91 (1909).

Sur le rôle du méthane dans la vie organique. Recueil d. Trav. chim. d. Pays-Bas **29,** 238 (1910).

Vetsplitsing door bakteriën. Versl. Kon. Akad. v. Wetensch. A'dam **19,** 689 (1910).

Fat-splitting by bacteria. Proc. Kon. Akad. v. Wetensch. A'dam **13,** 667 (1910).

Microben-lipase. Versl. Kon. Akad. v. Wetensch. A'dam **19,** 1263 (1911).

Lipase produced by microbes. Proc. Kon. Akad. v. Wetensch. A'dam **13,** 1200 (1911).

Thermo-tolerante lipase. Versl. Kon. Akad. v. Wetensch. A'dam **20,** 1263 (1911).

Thermo-tolerant lipase. Proc. Kon. Akad. v. Wetensch. A'dam **14,** 166 (1911).

Über fettspaltende Mikroben und deren Einfluss auf Molkereiprodukte und Margarine. Folia Microbiologica **1,** 199 (1912).

Oxydatie van petroleum, paraffine, paraffine-olie en benzine door microben. Versl. Kon. Akad. v. Wetensch. A'dam **21,** 1124 (1913).

Oxidation of petroleum, paraffin, paraffin-oil and benzine by microbes. Proc. Kon. Akad. v. Wetensch. A'dam **15,** 1145 (1913).

Benzin, Petroleum, Paraffinöl und Paraffin als Kohlenstoff- und Energiequelle für Mikroben. Centralbl. f. Bakt. II, **37,** 595 (1913).

Einfluss von Kolloiden auf mikrobiologische Prozesse. Centralbl. f. Bakt. II, **38,** 621 (1913).

Einfluss einiger Kolloide auf die Alkoholgärung. Folia Microbiologica **2,** 95 (1913).

Ueber reduzierende Eigenschaften der Essigbakterien. Folia Microbiologica **3,** 151 (1914).

Mit J. G. Fol, Die Zersetzung des Kautschuks durch Mikroben. Centralbl. f. Bakt. II, **40,** 87 (1914).

Invloed van eenige kolloiden op mikrobiologische processen. Chem. Weekbl. **11,** 42 (1914).

Het ontstaan en verdwijnen van mangani-verbindingen onder invloed van het microbenleven. Chem. Weekbl. **11,** 240 (1914).

Kolloidaal opgeloste en gelatineuse koolstof. Chem. Weekbl. **11,** 593 (1914).

Umwandlungen von Manganverbindungen unter dem Einfluss mikrobiologischer Prozesse. Centralbl. f. Bakt. II, **40,** 545 (1914).

Verslag over het onderzoek naar de oorzaken van het ontstaan van den stank der Haagsche grachten en aanwijzingen betreffende middelen ter verbetering. 's-Gravenhage, Gebr. Belinfante, 1914.

H. C. Jacobsen

Biologische beschouwingen over melk. 15e Jaarverslag Technol. Gezelschap, 25 (1906).

Ueber einen richtenden Einfluss beim Wachstum gewisser Bakterien in Gelatine. Centralbl. f. Bakt. II, **17,** 53 (1907).

Kulturversuche mit einigen niederen Volvocaceen. Zeitschr. f. Botanik **2,** 145 (1910).

Die Kulturbedingungen von *Haematococcus pluvialis.* Folia Microbiologica **1,** 163 (1912).

Die Oxydation von elementarem Schwefel durch Bakterien. Folia Microbiologica **1,** 487 (1912).

De samenstelling van het zetmeel. Chem. Weekbl. **10,** 552 (1913).

De oxydatie van zwavel tot zwavelzuur door bakteriën. Chem. Weekbl. **11,** 302 (1914).

Chemische reacties in colloide media. Chem. Weekbl. **11,** 588 (1914).

Die Oxydation von Schwefelwasserstoff durch Bakterien. Folia Microbiologica **3,** 155 (1914).

J. van der Leck

Aromabildende Bakterien in Milch. Centralbl. f. Bakt. II, **17,** 366, 480 und 647 (1907).

F. Liebert

Het afbreken van het urinezuur door bakteriën. Versl. Kon. Akad. v. Wetensch. A'dam **17,** 990 (1909).

The decomposition of uric acid by bacteria. Proc. Kon. Akad. v. Wetensch. A'dam **12,** 54 (1909).

A. Krainsky

Die Aktinomyceten und ihre Bedeutung in der Natur. Centralbl. f. Bakt. II, **41,** 649 (1914).

T. Folpmers

Tyrosinase, een mengsel van twee enzymen. Chem. Weekbl. **13,** 1282 (1916).

Tyrosinase, ein Gemenge von zwei Enzymen. Biochem. Z. **78,** 180 (1916).

Ontleding van koolhydraten door *Granulobacterium butylicum* Beijerinck. Tijdschr. v. Vergelijk. Geneesk. **6,** 33 (1920).

J. Smit

Studien über *Lactobacillus fermentum* (Beijerinck). Z. f. Gärungsphysiologie **5,** 273 (1916).

F. C. Gerretsen

Die Einwirkung des ultravioletten Lichtes auf die Leuchtbakterien. Centralbl. f. Bakt. II, **44,** 660 (1916). Cf. also: Über die Ursachen des Leuchtens der Leuchtbakterien. Centralbl. f. Bakt. II, **52,** 353 (1920).

P. van Steenberge

l'Autolyse de la levure et l'influence de ses produits de protéolyse sur le développement de la levure et des microbes lactiques. Ann. de l'Inst. Pasteur **31,** 601 (1917).

Les propriétés des microbes lactiques; leur classification. Ann. de l'Inst. Pasteur **34,** 803 (1920).

Appendix D.

List of doctor's theses wholly or largely prepared under BEIJERINCK's direction.

D. P. HOYER, Bijdrage tot de kennis van de azijnbacteriën. Delft, J. Waltman Jr., 1898.

C. J. J. VAN HALL, Bijdragen tot de kennis der bakterieele plantenziekten. Amsterdam, Coöp. Drukk. Ver. „Plantijn", 1902.

N. L. SÖHNGEN, Het ontstaan en verdwijnen van waterstof en methaan onder den invloed van het organische leven. Delft, J. Vis Jr., 1906.

A. RANT, De gummosis der *Amygdalaceae*. Amsterdam, J. H. de Bussy, 1906.

G. VAN ITERSON JR., Mathematische und mikroskopisch-anatomische Studien über Blattstellungen nebst Betrachtungen über den Schalenbau der Miliolinen. Jena, Gustav Fischer, 1907.

JAN SMIT, Bacteriologische en chemische onderzoekingen over de melkzuurgisting. Amsterdam, 1913.

J. A. HEYMANN, De voeding der oester. 's-Gravenhage, Mouton & Co., 1914.

F. C. GERRETSEN, Een onderzoek naar de nitrificatie en denitrificatie in tropische gronden. Epe, Stoomdrukkerij v.h. A. Hooiberg, 1921.

Appendix E.

Addresses made on September 30th, 1905 at the presentation of the LEEUWENHOEK Medal of the "Koninklijke Akademie van Wetenschappen te Amsterdam" to BEIJERINCK. *)

De Heer WENT houdt de volgende toespraak:

Hooggeachte Heer BEIJERINCK.

Aan mij valt heden het voorrecht ten deel, U te mogen toespreken nu U de LEEUWENHOEK-medaille zal worden uitgereikt. De Commissie, die over de toewijzing had te beslissen (waarvan de Voorzitter tot zijn leedwezen tengevolge van een Regeeringsopdracht afwezig is) heeft mij die taak opgedragen, daar ik het eenige botanische lid in haar midden ben, maar ik heb die opdracht met vreugde aanvaard, ook omdat mijn werk mij dikwijls van meer nabij met het Uwe heeft doen kennismaken.

Toen onze Commissie zich voor de vraag gesteld zag uit te maken, wie in de laatste tien jaren het meest had bijgedragen tot de vermeerdering van de kennis der mikroskopisch kleine wezens, bleek al spoedig, dat haar taak niet zoo heel gemakkelijk was. Langzamerhand toch is het veld van studie der mikroskopisch kleine organismen zoo groot geworden, dat het voor één enkel mensch niet mogelijk is dit geheel te overzien, zoodat ook daar een sterke specialiseering is waar te nemen; het valt den botanicus daardoor moeilijk te oordeelen over de waarde van bacteriologisch medische onderzoekingen, den bacterioloog over zoölogische waarnemingen en zoo omgekeerd. Toch waren wij het er al spoedig over eens, dat, waar helaas bij zoo velen de neiging bestaat de mikro-organismen eenigszins te plaatsen buiten de overige levende wezens, zeker voor de toekenning der medaille in de eerste plaats het oog gevestigd zou moeten worden op die onderzoekingen, welke een meer algemeene beteekenis hebben, die een helderder licht doen vallen op algemeene biologische vraagstukken en toen was het natuurlijk, dat al dadelijk Uw naam genoemd werd en dat het ons voorkwam, dat niemand meer dan Gij aanspraak mocht maken op de toekenning der LEEUWENHOEK-medaille in het jaar 1905.

Niet alleen LEEUWENHOEK, maar ook onderzoekers, die veel later leefden, hebben er zeker niet van gedroomd, dat de studie dier mikroskopisch kleine wezens ons in vele opzichten zou kunnen leiden tot een betere kennis van tal van levensproblemen, die men bij deze organismen in hun eenvoudigsten vorm aantreft, ja ik geloof zelfs te mogen zeggen, dat bij degenen, die het fonds voor de LEEUWENHOEK-medaille tot stand brachten, dergelijke denkbeelden nog niet bestonden. Hoezeer is in 30 jaar de stand van de wetenschap in dat opzicht veranderd! Maar tot degenen, die door hun onderzoekingen hier nieuwe inzichten deden ontstaan, behoort Gij zeker in de allereerste plaats. Niet alleen in de laatste tien jaren, maar reeds lang te voren, sedert Gij Uw woonplaats verlegd hebt naar Delft, de stad van LEEUWENHOEK, zijt Gij bezig geweest met de studie der mikroben. Toch is er een tijd geweest, dat daarbij vooral vraagstukken van de praktijk en pas in de tweede plaats zuiver wetenschappelijke vragen uw aandacht hadden bezig te houden. Dit werd anders sedert Gij nu juist 10 jaar geleden als hoogleeraar zijt opgetreden aan de Polytechnische School, thans Technische Hoogeschool. In de redevoering, waarmede Gij toenmaals Uw ambt aanvaard hebt „De biologische Wetenschap en de Bacteriologie", werd door U wel is waar ook gewezen op de beteekenis van de studie der mikrobiologie voor de praktijk, maar toch werd hier nadruk gelegd op het groote belang, dat er in gelegen is om algemeene biologische vraagstukken te bestudeeren bij de eencellige organismen, vooral omdat men hier mist de complicatie van een arbeidsverdeeling tusschen verschillende cellen, omdat in het algemeen tal van problemen zich hier veel gemakkelijker laten stellen en men hier zeker het allereerst tot hun oplossing zal kunnen geraken. In die richting hebt Gij zelf ook altijd gewerkt en zonder dat het in mijn

*) Reprinted from Versl. Kon. Akad. v. Wet. Amsterdam **14,** 203, 1905.

bedoeling ligt hier een opsomming te geven van het vele, dat door U op mikrobiologisch gebied gevonden is, zou ik toch enkele der meest in het oog vallende van uw ontdekkingen der laatste 10 jaren kort willen herdenken.

In de eerste plaats dan de onderzoekingen over assimilatie van vrije stikstof. Reeds vroeger was door U een zeer belangrijke bijdrage geleverd tot de kennis der organismen, die in symbiose met Leguminosen stikstof assimileeren; thans hebt Gij ook Uw aandacht gewijd aan het stikstofvraagstuk in veel ruimeren zin en dit is aanleiding geweest tot de ontdekking van soorten van het geslacht *Azotobacter*, vooral van *Azotobacter chroococcum*. Was tot nu toe alleen de anaërobe *Clostridium Pasteurianum* beschreven als in staat vrije stikstof te assimileeren, Uw onderzoekingen maakten ons bekend met aërobe bacteriën, die ditzelfde vermogen, waarbij in het midden gelaten kan worden of zij hiertoe alleen in staat zijn, dan wel in symbiose met andere mikroben. Was door U zelf reeds gewezen op de groote verspreiding van *Azotobacter* in de natuur, latere onderzoekingen hebben dit nog meer bevestigd en de overtuiging veld doen winnen, dat, wat betreft de vorming van stikstofverbindingen uit vrije stikstof deze organismen zeker een zeer belangrijke rol spelen. Dat dit resultaat door U bereikt werd, terwijl vroeger zooveel onderzoekers vergeefs gezocht hadden, moet wel vooral toegeschreven worden aan de genialiteit van uw onderzoekingsmethoden, waarbij zooveel mogelijk de omstandigheden in de natuur werden nagebootst en waarbij pas in de laatste instantie met reinkulturen gewerkt werd. Daarbij kwam in de tweede plaats het gelukkige denkbeeld van het bestaan van oligonitrophile mikroben, die dus ook wel stikstofverbindingen als voedsel kunnen bezigen, maar alleen wanneer deze in zeer verdunden toestand gegeven worden.

Ik denk verder aan Uw proefnemingen met Bacteriën, die koolzuur als koolstofbron in het donker kunnen gebruiken. Werd reeds vroeger door U aangetoond, dat men op een dwaalspoor zou kunnen komen bij *Bacillus oligocarbophilus*, daar deze leven kan ten koste van de zeer geringe hoeveelheden organische stoffen, die in de laboratoriumslucht voorkomen, verleden jaar werden de proeven van Nathansohn door U uitgebreid, waardoor blijkt dat koolzuur gereduceerd kan worden door mikroben die hun energie verkrijgen hetzij uit zwavelwaterstof, thiosulfaat of tetrathionaat of door denitrificatie met vrije zwavel.

De methode om van massakulturen uit te gaan, waarbij de omstandigheden zoo gekozen werden, dat slechts die mikroben zich ontwikkelen, die aan bepaalde levensvoorwaarden geadapteerd zijn, heeft U niet alleen hier, maar ook in andere gevallen tot belangrijke resultaten gevoerd. Ik denk aan Uw proeven met Ureumbacteriën, aan die over boterzuurgisting, over sulfaatreduceerende organismen, vooral aan die over anaërobiose. Hier geldt het een derde hoogst belangrijk vraagstuk, aan welks oplossing Gij meewerkt. Pasteur had onze denkbeelden omtrent ademhaling een radicale wijziging doen ondergaan door zijn ontdekking van anaërobe organismen. Met behulp van de fraaie methode der sedimentfiguren bij bewegelijke bacteriën kon door U aangetoond worden, dat verschillende mikroben zeer verschillende zuurstofspanningen opzoeken, dat zij een zeer verschillende behoefte aan vrije zuurstof hebben. Uw voortgezette onderzoekingen voerden U ten slotte tot de voorstelling, dat ook de zoogenaamd obligaat anaërobe organismen vrije zuurstof noodig hebben, zij het dan ook zeer weinig, zoodat volgens U beter gesproken wordt van mikroaërophilen.

Wanneer hier over ademhaling gesproken wordt, denkt natuurlijk ieder ook dadelijk aan Uw studiën over lichtende bacteriën, die zulk een aantal verrassende nieuwe feiten leerden kennen; deze zijn wel is waar niet afkomstig van de laatste tien jaar, maar Gij hebt de toen het eerst gebezigde methode, de auxanographische, ook later nog herhaaldelijk toegepast met schitterend succes. Ik wijs daarbij b.v. op uw onderzoek over de glukase en over het voorkomen daarvan, over het indigo-enzym, over sulfaatreduceerende Bacteriën en zoo vele andere onderzoekingen op het gebied van stofwisselingsprocessen. Hoe belangrijk deze ook zijn, ik zal er hier niet verder op ingaan, om even de aandacht te vragen voor eenige van Uw verhandelingen, die zich op een geheel ander gebied bewegen.

Ik bedoel die, welke betrekking hebben op de veranderlijkheid van mikroben. Steeds werd Uw geest aangetrokken door de studie der erfelijkheidsproblemen; ik behoef slechts te noemen: Uw galstudiën, Uw onderzoek van *Cytisus Adami*. Het is dan ook begrijpelijk, dat Gij voor deze problemen bij de mikroben naar een oplosssing gezocht hebt. In Uw reeds genoemde redevoering hebt Gij er op gewezen, dat in de eerste plaats bij mikro-organismen in zeer korten tijd beschikt kan worden over geheele reeksen van generaties, en dat ten tweede wijziging van uitwendige omstandigheden diepere veranderingen teweegbrengt van de erfelijke eigenschappen, dan men dit ergens bij de hoogere organismen heeft waargenomen. Maar Gij hebt zelf onderzoekingen verricht op dit gebied, b.v. over het verlies van het sporevormend vermogen bij alcoholgisten, maar vooral denk ik daarbij aan de mededeeling hier in deze Akademie door U gedaan op 27 October 1900 over verschillende vormen van erfelijke variatie bij mikroben en aan uw verhandeling van verleden jaar over „*Chlorella variegata,* ein bunter Mikrobe", een verhandeling waarvan de titel reeds wijst op het merkwaardige van den inhoud. In dit laatste geval hieldt Gij U bezig met groene organismen en dit geeft mij aanleiding nog te wijzen op Uw groote verdiensten, wat de studie der lagere wieren betreft.

Tot voor korten tijd was een studie der mikroskopisch kleine wieren niet wel mogelijk, zoodra men hun eigenschappen wilde onderzoeken geheel onafhankelijk van andere levende wezens. Immers het was niet mogelijk ze in reinkultuur te kweeken; niet alleen wist men weinig of niets van hun physiologische eigenschappen, maar zelfs hun ontwikkelingsgeschiedenis was niet voldoende bekend en tal van duistere punten moesten hier opgehelderd worden. Gij hebt het eerst de mogelijkheid aangetoond van kultuur van groene mikroben op soortgelijke voedingsbodems als de niet groene, eerst van *Chlorella vulgaris,* later van *Pleurococcus vulgaris* en andere, zelfs van *Diatomeae.* Schijnbaar kleine onderzoekingen, maar die den grondslag vormen voor een omwenteling in de studie der lagere Algen, die thans op dezelfde wijze aangevat wordt als met Bacteriën een 25-tal jaren geleden het geval is geweest.

Zooals ik zooeven al zeide, was het niet mijn doel hier een opsomming te geven van al uw onderzoekingen; ik heb slechts op enkele van de meest belangrijke een oogenblik het licht doen vallen, maar ik ga er andere van veel beteekenis voorbij, b.v. die over azijngist, over zwavelwaterstofvorming en over het geslacht *Aërobacter,* over de oorzaak der mozaiekziekte van de tabak en nog zooveel meer. Trouwens het is uilen naar Athene dragen, wanneer ik er hier op deze plaats over spreek, immers de meeste van Uw vele onderzoekingen zijn het eerst in de werken van deze Akademie gepubliceerd, vele na een mondelinge voordracht er over. Juist de algemeene beteekenis uwer proefnemingen maakte, dat Gij hier steeds een zeer aandachtig gehoor hadt.

Wij verheugen ons er over, dat het een Nederlander is, aan wie de Nederlandsche LEEUWENHOEK-medaille ditmaal gegeven wordt, terwijl bij vorige gelegenheden aan vreemdelingen die eer te beurt viel. Nog één verschil is er; toen waren het, zooals meestal het geval is bij dergelijke eerbewijzen, mannen, die hun levenstaak vervuld hadden, van wie niet veel meer op wetenschappelijk gebied verwacht kon worden. Gij bevindt U in de kracht van Uw leven, Uw wetenschappelijke productiviteit is veeleer stijgende dan afnemende en wij mogen dan ook wel de hoop uitspreken, dat nog veel belangrijke ontdekkingen op mikrobiologisch gebied door U gedaan zullen worden. Wanneer het mij vergund is daarbij een wensch te uiten, dan weet ik, dat velen met mij gaarne eens een samenvatting van Uwe hand zouden zien van Uw denkbeelden over de biologie der mikro-organismen. Er ligt in Uw verschillende verhandelingen zulk een schat van oorspronkelijke denkbeelden en bijzondere opvattingen dikwijls in enkele zinnen begraven, dat een dergelijke samenvatting zeker met spanning tegemoet zou worden gezien. Er zou ook uit blijken, hoeveel van de tegenwoordige voorstellingen op mikrobiologisch gebied wij eigenlijk aan U te danken hebben; dit is veel meer, dan menigeen weet, die slechts oppervlakkig van Uw werk kennis genomen heeft. Ook daardoor zal Uw naam steeds genoemd worden onder de Nederlanders, die belangrijk bijgedragen hebben tot vermeerdering van onze kennis op natuurhistorisch gebied, waardoor Gij de waardige nakomeling zijt van een INGENHOUSZ, een SWAMMERDAM, een LEEUWENHOEK.

De Voorzitter dankt den Heer WENT voor het uitgebrachte verslag en overhandigt de gouden medaille aan den Heer BEIJERINCK, waarna deze, het woord verkregen hebbende, het volgende zegt:

Mijnheer de Voorzitter, Mijnheer WENT!

Ontvangt mijn dank voor Uwe hartelijke woorden, die zoo ondubbelzinnig bewijzen, dat de richting, waarin ik de Mikrobiologie beoefen, de sympathie wegdraagt van de beste beoordeelaars.

Die richting is kort te omschrijven als het onderzoek van de Oekologie der mikroben, dat is van het verband tusschen bepaalde levensvoorwaarden en bepaalde levensvormen die daaraan beantwoorden. Daar het mijn overtuiging is, dat deze bij den tegenwoordigen stand der wetenschap de meest noodzakelijke en meest vruchtbare richting is om orde te brengen in onze kennis aangaande dat deel van het natuurlijke stelsel, dat de laagste grens omvat van de organische wereld, en dat ons aanhoudend het groote vraagstuk naar den oorsprong van het leven zelve in scherpe trekken voor oogen stelt, is het mij tot groote voldoening, dat de Akademie blijkbaar de beoefening daarvan in den beoefenaar wil bekronen.

In experimenteelen zin geeft de oekologische opvatting der Mikrobiologie, in twee elkander aanvullende richtingen aanleiding tot een eindeloos getal van proeven, namelijk eenerzijds tot het opsporen van de levensvoorwaarden van reeds door een of andere omstandigheid of door het toeval bekend geworden mikroben, en anderzijds tot de ontdekking van levende wezens, welke bij vooraf vastgestelde levensvoorwaarden verschijnen, hetzij omdat *zij* alleen daarbij kunnen bestaan, of omdat juist zij bij die invloeden de sterksten zijn en hun medestanders overwinnen. Vooral deze laatste methode van onderzoek, die eigenlijk niets anders is dan de ruime toepassing van wat tegenwoordig veelal de elektieve kultuurmethode genoemd wordt, is vruchtbaar en echt wetenschappelijk, en het is niet te veel om te zeggen, dat de Algemeene Mikrobiologie vooral daaraan haren veelzijdigen en verrassenden vooruitgang te danken heeft. Maar ofschoon reeds LEEUWENHOEK voor meer dan twee eeuwen bij sommige van zijn onderzoekingen deze zijde der Mikro-oekologie in praktijk bracht, en PASTEUR daardoor geleid de meeste zijner groote ontdekkingen heeft kunnen doen, is het getal van bewuste beoefenaren daarvan tot nu toe slechts zeer gering gebleven, en ik gevoel, dat ik zeker daartoe mag gerekend worden door den lust die in mij is om bij te dragen tot het grootsche werk, dat op dit gebied te volbrengen valt.

Maar de verdieping der inzichten in een zoo subtielen en moeilijken tak van kennis als de Mikrobiologie schrijdt uiterst langzaam voort, en daarom moet een gebeurtenis als deze, naast voldoening, nog gedachten van anderen aard opwekken. Komende, wanneer het hoogtepunt van het leven bereikt of voorbij is, de blik in het tegenwoordige en de toekomst het helderste is geworden, kunnen er geen illusies meer bestaan aangaande de verhouding van de nieuw gevonden wetenschappelijke feiten tot de afmetingen van den oceaan der waarheid.

Toch zal de herinnering aan dit bewijs van waardeering,toegewezen door een kring als deze, mij als Nederlandsch geleerde, bij het klimmen der jaren, en wanneer de wetenschap zal ophouden haar loon in zich zelf te dragen, ongetwijfeld de voortzetting van de bewerking van het gekozen arbeidsveld veraangenamen en verlichten, en wèl dus heb ik reden, u mijne heeren, Voorzitter en Leden der Akademie, mijne gevoelens van groote erkentelijkheid voor dit onvergetelijke oogenblik aan te bieden.

Appendix F.

Article published by Professor S. HOOGEWERFF on the occasion of the silver jubilee of BEIJERINCK's professorship. *)

PROF. IR. DR. M. W. BEIJERINCK, 1895—1920

Het 25-jarig jubileum van prof. BEIJERINCK op 1 Juli 1920 mag in *De Ingenieur* niet onbesproken blijven. Immers, hoewel hij in de eerste plaats botanicus is, houdt zijn werkzaamheid in haar aard en ontplooiing zoo nauw verband met de techniek in ruimen zin, dat ook in dit tijdschrift BEIJERINCK's 25-jarig professoraat aan de P. S. en T. H. met een kort woord herdacht moge worden. Dit geschiede zonder vooruit te loopen op uitvoerige beschouwingen over zijn wetenschappelijke verdiensten en zijn werken, beschouwingen, welke zeker van bevoegder hand dan de mijne zullen worden gegeven, als hem, ter gelegenheid van zijn 70-sten geboortedag, in Maart 1921, op meer afdoende wijze en in ruimen kring de hem toekomende hulde zal worden gebracht, waartoe de plannen in voorbereiding zijn.

BEIJERINCK werd geboren te Amsterdam 16 Maart 1851. Hij doorliep de H.B.S. 5-j. c. en studeerde daarna aan de Polytechnische School voor technoloog, tegelijkertijd met J. H. VAN 'T HOFF, met wien hij te Delft nauwe vriendschapsbanden sloot en dezelfde kamers bewoonde. In 1872 verkreeg hij het diploma van technoloog, om daarna onder SURINGAR te Leiden botanie te gaan studeeren[1]). Reeds vóór zijn promotie vinden wij hem aan de Landbouwschool te Wageningen als docent in dat vak. In 1877 behaalde hij den graad van doctor in de wis- en natuurkunde, na verdediging van zijn proefschrift: ,,Bijdrage tot de morphologie der plantengallen''. De talentvolle jeugdige botanicus, die tevens bleek een uitnemend docent te zijn, zette zijn wetenschappelijke onderzoekingen met ijver voort, en reeds in 1884 werd hij tot lid der Kon. Akademie van Wetenschappen (Wis- en Natuurkundige Afd.) gekozen, in wier werken reeds voordien tal van belangrijke verhandelingen van zijn hand waren verschenen. Alles wees er op, dat voor hem de academische loopbaan zich spoedig zou openen.

Doch door de breede opvattingen en den juisten blik van J. C. VAN MARKEN, directeur der Ned. Gist- en Spiritusfabriek te Delft, werd reeds in 1885 BEIJERINCK verbonden aan die toen nog in de jaren der kinderziekten verkeerende onderneming; in een tijd dus toen wetenschappelijke hulpkrachten in de Nederlandsche industrie nog nagenoeg onbekend waren. In een voor hem gesticht bacteriologisch laboratorium kon hij zijn groote gaven geheel wijden aan de bevordering der gistingsindustrie, waarbij hem echter volledige vrijheid in de keuze zijner onderzoekingen werd gelaten.

Het zij mij vergund hier uit de *Fabrieksbode* van 11 April 1885 de nobele woorden aan te halen, waarmede VAN MARKEN, in een uitnemend geschreven artikel — populair in den goeden zin — getiteld: ,,Bacteriologie'', BEIJERINCK's komst aan de Gist- en Spiritusfabriek en de daarvan te koesteren verwachtingen vermeldt.

„Een jong geleerde, maar die zijn sporen op het gebied der natuurwetenschap reeds heeft verdiend, de heer dr. M. W. BEIJERINCK, heeft het niet beneden zijn wetenschappelijke waardigheid geacht, de taak van een VON MOLTKE in ons nijverheidsbedrijf te aanvaarden. Hij heeft gemeend hier een bij uitnemendheid rijk veld van onderzoek te vinden. Hij verwacht van de navorsching der geheimen, die hier verborgen liggen, hoogere bevrediging — de bevrediging van den ernstigen natuuronderzoeker — dan enkel die van het stoffelijk voordeel, dat wij als een gevolg van zijn arbeid voor onze onderneming mogelijk achten en waarop wij hopen.''

[1]) BEIJERINCK vormde met J. H. van 'T HOFF en A. A. W. HUBRECHT het drietal, ten wiens gunste THORBECKE het veel bestreden besluit wist uit te lokken tot toelating tot de studie in de wis- en natuurkunde aan de universiteit, ruim 30 jaar vóórdat de wet-LIMBURG voor het einddiploma H.B.S. 5-j.c. die bevoegdheid erkende. De geschiedenis heeft bewezen, dat het drietal het gunstbetoon waard was.

*) Reprinted from "De Ingenieur" **35**, 482, 1920.

Na vermeld te hebben, dat voor hem een laboratorium wordt gebouwd, „afgescheiden van het gewoel en gedruisch in onzen rumoerigen bijenkorf en voorzien van de meest volkomen mikroskopen en van andere wetenschappelijke werktuigen en inrichtingen", gaat hij voort:

„Zullen de onderzoekingen practische vruchten voor onze onderneming afwerpen? De heer BEIJERINCK is bescheiden en wetenschappelijk genoeg om dit vraagteeken voorloopig onbeantwoord te laten. Uitdrukkelijk heeft hij dit verklaard, toen hij op mijn wensch zich bereid verklaarde de taak te aanvaarden. Wat weten wij nog, na zoovele eeuwen van onderzoek en ontwikkeling, van het raadsel dat leven wordt genoemd? De meest uitstekende geneeskundige staat menigmaal schouderophalend aan het ziekbed van den mensch, die wat hij voelt en waar hij lijdt, kan mededeelen en aanwijzen. En hier hebben wij te doen met levende wezens, die, met behulp van de meest volkomen instrumenten, nog nauwlijks zijn waar te nemen.

Hoe het zij, de komst van een geleerde als dr. BEIJERINCK is in meer dan één opzicht een belangrijk feit, dat in onzen kring hooge waardeering verdient. Ik wensch volstrekt geen overdreven verwachtingen van zijn werkzaamheid in en voor onze fabriek op te wekken. Maar wel ben ik overtuigd, dat ernstige wetenschappelijke arbeid op het gebied der bacteriologie te eeniger tijd — over een jaar, vijf, tien jaren misschien, wij hebben geloof in de wetenschap en haasten haar niet — een enkel straaltje van licht zal werpen in de duisternis van het gistingsbedrijf en wellicht onberekenbare voordeelen aan onze onderneming zal kunnen brengen."

Al zijn, uit den aard der zaak, omtrent de diensten, welke BEIJERINCK aan de Gist- en Spiritusfabriek heeft bewezen, de bijzonderheden niet algemeen bekend, zoo is het toch wel haast overbodig er hier op te wijzen, dat die verwachtingen op schitterende wijze zijn verwezenlijkt.

Weinig bevroedde VAN MARKEN toen wel, dat hij, door BEIJERINCK van Wageningen naar Delft te roepen, tevens de aanleiding schiep, dat deze na eenige jaren gewonnen zou worden voor de school, waaraan VAN MARKEN zelf zijn opleiding had genoten. Spoedig toch na BEIJERINCK's komst te Delft, rijpte bij A. C. OUDEMANS het denkbeeld om hem uit te noodigen een cursus te geven in de technische botanie en het was op aandrang van dien toenmaligen directeur der P. S., dat de Regeering aan dr. M. W. BEIJERINCK vergunning verleende „om buiten bezwaar van 's Lands schatkist aan de Polytechnische School lessen te geven over onderwerpen van plantkundigen aard, met de nijverheid in verband staande". Onze jubilaris is dus de eerste privaat-docent aan de P.S. geweest.

Zij, die in de eenvoudige college-kamer voor scheikunde aan de Westvest bedoelde voordrachten, die eenmaal 's weeks werden gegeven, hebben bijgewoond — en ik behoor tot die gelukkigen —, herinneren zich naast den meesleependen vorm, de groote helderheid en het belangwekkende van dat onderwijs.

Alras was de Raad van Bestuur dan ook overtuigd, dat het van het hoogste belang was om een zoodanige kracht voor goed en in ruimeren werkkring aan de Polytechnische School te verbinden en dit in de eerste plaats ter aanvulling van de opleiding der a.s. technologen. Aanvankelijk hadden de daartoe bij de Regeering aangewende pogingen geen gunstig resultaat. Doch, dank zij ook de medewerking van enkele leden der Tweede Kamer — ik noem MEES en VAN DE VELDE — en den aandrang uit nijverheidskringen, werd bij K.B. van 24 Juni 1895 dr. M. W. BEIJERINCK benoemd tot hoogleeraar aan de Polytechnische School om onderwijs te geven in biologie en bacteriologie. Hem werd voorloopig aangewezen het oude postkantoor aan het Oude Delft, waar hij een tijdelijk laboratorium, later bij het microchemisch onderwijs in gebruik genomen, inrichtte, en op 6 September van genoemd jaar opende hij zijn lessen met een rede: „De biologische wetenschap en de bacteriologie".

Noode verliet BEIJERINCK de Gist- en Spiritusfabriek, waar hij zich geheel aan onderzoek had kunnen wijden. In die betrekking werd hij opgevolgd door H. P. WIJSMAN, die aldaar reeds als BEIJERINCK's assistent werkzaam was geweest en onder diens leiding zijn proefschrift: „De diastase beschouwd als mengsel van maltase en dextrinase" had bewerkt, dat menig punt van aanraking met BEIJERINCK's eigen onderzoekingen heeft en waarnemingen bevat, waaruit later een belangrijke vooruitgang op het gebied der nijverheid zou voortkomen.

Was de beperkte inrichting op het Oude Delft voor BEIJERINCK geen beletsel om daar dadelijk met alle kracht in zijn nieuwe betrekking werkzaam te zijn, zoo was die behuizing toch van den aanvang af slechts als een provisorische bedoeld. Een nieuw laboratorium, met proeftuin, werd, grootendeels naar zijn aanwijzingen, in de Nieuwe Laan gebouwd en reeds in 1897 kon hij met een redevoering: „Het bacteriologisch laboratorium der Polytechnische School" dat laboratorium openen, eenige jaren geleden iets vergroot en thans nauwelijks meer voldoende ruimte aanbiedend.

Nagenoeg een kwart eeuw is hij daar onvermoeid werkzaam geweest aan zijn eigen wetenschappelijke onderzoekingen, waarvan het meerendeel ook een technischen kant bezit, en aan de opleiding zijner leerlingen.

Staan wij bij beide vormen zijner werkzaamheid nog een oogenblik stil.

Het is, zooals ik in den aanvang motiveerde, niet mijn bedoeling hier een opsomming te geven zijner talrijke wetenschappelijke onderzoekingen, die in de *Verslagen en Verhandelingen der Kon. Akademie van Wetenschappen*, in de *Archives Néerlandaises*, in de *Botanische Zeitung*, in het *Centralblatt für Bakteriologie*, in de *Folia Microbiologica* e.a.m. zijn verschenen. Maar ik wil toch, om ook den lezer van dit tijdschrift eenig denkbeeld te geven van den omvang en beteekenis van BEIJERINCK's werkzaamheid als natuuronderzoeker, althans vermelden, dat in de eerste jaren na de studie over de gallen — waarover zijn proefschrift handelt en dat vrij spoedig door zeer belangrijke publicaties over hetzelfde onderwerp werd gevolgd — verschillende verhandelingen van zijn hand over botanische onderwerpen verschenen, waaronder verscheidene van phytopathologischen aard. Daarop (1888) bestudeert hij de uitwassen aan de wortels van de Papilionaceeën en de daarbij werkzame bacteriën, en publiceert later (1902) met VAN DELDEN een onderzoek over de assimilatie van vrije stikstof door bacteriën buiten medewerking van de plant.

Inmiddels verbetert hij verschillende bacteriologische onderzoekingsmethoden of voegt aan de bestaande nieuwe toe en wijdt zich tevens aan een studie van de lichtgevende bacteriën; een onderzoek, dat hem ook van een algemeen natuurwetenschappelijk standpunt aantrok, hem lang bezig hield en voor WIJSMAN's proefschrift en ook voor andere onderzoekingen van BEIJERINCK resultaten heeft afgeworpen.

De nitrificatie-verschijnselen in den bodem, de sulfaat-reductie — ook die in de openbare wateren — en haar gevolgen, verschillende studies over alcoholgisting, een uitvoerige verhandeling over de butylalcoholgisting en het butylferment, over een contagium vivum fluidum, als oorzaak van de mozaïkziekte bij de tabak, belangrijke onderzoekingen over de indigovorming uit weede en over de indigo-fermentatie, de melkzuurfermenten in de nijverheid en over de melkzuurgisting in melk, over de bacterie, welke bij het rooten van het vlas werkzaam is (met VAN DELDEN in 1904) — ziedaar eenige grepen uit den rijken schat zijner onderzoekingen, waarbij ik, als chemicus, er wel onvermeld laat, die de botanicus of bacterioloog juist allerbelangrijkst zal achten.

Bovendien heeft BEIJERINCK herhaaldelijk op congressen of vergaderingen samenvattende voordrachten over microbiologische onderwerpen gehouden; ik noem zijn voordracht in 1904 in de vergadering van de Hollandsche Maatschappij der Wetenschappen: „De invloed der mikroben op de vruchtbaarheid van den grond en op den groei der hoogere planten".

Doch mijn schets zou geheel onvolledig zijn, wanneer ik ten slotte niet wees op den invloed, dien BEIJERINCK als docent en niet minder als leider van zijn laboratorium, in zijn 25-jarige werkzaamheid als hoogleeraar, op zijn leerlingen heeft uitgeoefend door zijn bezielend onderwijs. Dat adjectief is geen gelegenheidsvorm; doch naar mijn overtuiging en ervaring de juiste omschrijving van zijn onderricht. Altijd wist hij de volle aandacht zijner leerlingen te boeien bij de mededeeling van zijn rijke kennis, als hij hen inwijdde in de subtiele vraagstukken der microbiologie, hun de methoden van het microbiologisch onderzoek onderwees, de besten zijner leerlingen tot zelfstandige onderzoekers vormde en bovendien van allen het inzicht in de natuur en in de bedrijven, waar microbiologische werkwijzen worden toegepast, verruimde. Te verwonderen is het dan ook niet, dat behalve studenten-technologen (de microbiologie behoort aan de T.H. tot de facultatieve vakken) ook a.s. industrieelen en beoefenaren der microbiologie ook uit onze Koloniën en uit het buitenland in den loop der jaren BEIJERINCK's leiding zochten en in zijn laboratorium werkzaam waren. Niet licht zal men den invloed overschatten door zijn onderricht uitgeoefend. Is het moeilijk deze in alle bijzonderheden na te gaan, ik meen te kunnen volstaan met er op te wijzen, dat thans in Nederland een drietal hoogleeraren werkzaam zijn, die hun opleiding voor een belangrijk gedeelte door BEIJERINCK hebben ontvangen en waarvan twee zijn oud-assistenten zijn, die ook hun proefschriften onder zijn leiding bewerkten; dat de drinkwatervoorziening in onze twee grootste gemeenten geleid wordt door zijn oud-leerlingen; dat de directeuren der beide bacteriologische afdeelingen aan de Rijkslandbouwproefstations voor landbouwkundig onderzoek BEIJERINCK's onderricht genoten, eveneens de directeur van het Rijksinstituut voor Hydrog. Visscherijonderzoek; en dat ook in onze koloniën en in de Nederlandsche nijverheid enkele zijner leerlingen belangrijke functies uitoefenen.

De talentvolle jubilaris, wiens naam aan het hoofd van dit artikel werd geplaatst, zal op een welbesteed leven terugzien als hij over eenige maanden, door de wet gedwongen, zijn betrekking als hoogleeraar moet neerleggen. Moge het hem gegeven zijn aan zijn lievelingswetenschap nog lang zijn krachten te kunnen wijden. Daarover zullen zich ook zijn vrienden verheugen, die aan het samenzijn met hem zoo menige opwekking, ook in wetenschappelijk opzicht, te danken hebben.

S. HOOGEWERFF

Appendix G.

Address delivered by Professor G. van Iterson Jr. on March 16th 1921 on the occasion of the seventieth anniversary of Beijerinck's birthday. *)

JUBILEUM PROFESSOR BEIJERINCK 1851—1921

In de laatste periode van zijn rijke wetenschappelijke loopbaan legde de grondlegger onzer moderne bacteriologie, Louis Pasteur, zich in hoofdzaak toe op de studie der infectieziekten van de dieren en den mensch. De resultaten, die hij wist te verkrijgen bij de bestrijding van het miltvuur en vooral van de hondsdolheid werden door de geheele wereld — en terecht — met bewondering aanschouwd. Toen nu de Duitsche arts Robert Koch in 1882 de tuberkelbacil als bewerker der tuberculose en in 1884 de kommabacil als oorzaak der cholera kon aanwijzen en deze vondsten weldra door de ontdekking van vele andere mikroskopische bewerkers van menschelijke en dierlijke ziekten werden gevolgd, begon zich langzamerhand het begrip „bacteriologie" vast te koppelen aan dat van „ziekteleer".

Men zag daarbij geheel over het hoofd, dat de eerste onderzoekingen van Pasteur op bacteriologisch gebied, onderwerpen van geheel anderen aard betroffen. In 1857 had hij aan de Akademie van Wetenschappen te Parijs een verhandeling over melkzuurgisting overhandigd, weldra gevolgd door onderzoekingen over het alcoholferment, de boterzuurgisting en de azijnbacteriën. Maar dit alles trok alleen in zeer beperkten kring de aandacht. Slechts langzaam is in latere jaren het begrip doorgedrongen, dat naast de leer der pathogene mikroben een volkomen gelijkwaardige studie-richting staat, die de rol der mikro-organismen in de huishouding der natuur tot object van onderzoek heeft. Maar geenszins is het nog van voldoende bekendheid, hoe veelomvattend deze studierichting is, hoe de *gezonde* mensch aan mikrobenwerkingen zijn bestaansmogelijkheden dankt, hoe de stof- en energieomzettingen, door de mikro-organismen in de natuur teweeggebracht, qualitatief en quantitatief niet onderdoen voor die, welke in het planten- en het dierenrijk tezamen verloopen.

Wij weten thans, hoe de voortdurende kringloop, welke de koolstof, zuurstof, waterstof, stikstof en zwavel moeten doorloopen, om het telkens zich vernieuwende leven de noodzakelijke bouwstoffen en de onmisbare energievormen toe te voeren, onafscheidelijk verbonden is aan de werkingen van de algemeen om ons heen voorkomende, laagst georganiseerde wezens. Sedert het koolzuurgehalte van onzen atmospheer aan nauwkeurige bepalingen werd onderworpen — dat is reeds meer dan 100 jaren — is daarin geen verandering geconstateerd. Dat dit gehalte in den loop van eeuwen constant moet zijn gebleven, is ook wel indirect te besluiten, omdat reeds een betrekkelijk kleine verandering in dat lage koolzuurgehalte het geheele beeld der vegetatie op onze aardoppervlakte zou wijzigen, ongetwijfeld een belangrijke verandering in de oogstopbrengsten zou veroorzaken en zeer waarschijnlijk zelfs een groote wijziging der klimaten zou meebrengen. Die opvallende gelijkmatigheid in het voor de planten onmisbare koolzuur — de koolstofbron, waaruit alle leven op aarde put — is slechts mogelijk door de activiteit van mikroben, die naar schatting jaarlijks meer dan honderd billioen K.G. organische materie afbreken tot eenvoudige stoffen en de koolstof weer als koolzuur aan de atmospheer teruggeven.

Dat de planten op het grootste deel van onze aardoppervlakte ook in den bodem de zeer speciale en engbegrensde mogelijkheden tot ontwikkeling hunner wortels en tot opname der, naast de koolstof noodige, elementen vinden, is eveneens uitsluitend aan een samenstel van zeer merkwaardige mikrobenprocessen te danken. Wij hebben langzamerhand de zekerheid gekregen, dat de verscheidenheid der chemische omzettingen, die in den bodem voortdurend naast elkander onder den invloed van mi-

---

*) Reprinted from Vakblad voor Biologen **2,** 1921 (Special number); a German translation has appeared in Zeitschr. f. techn. Biol. **9,** 235, 1921.

kroben verloopen, grooter is dan die, welke in de uitgebreidste der chemische fabrieken in gang zijn.

Naarmate zich de kennis van deze rol der lagere organismen uitbreidde, leerde men hunne werkingen ook steeds meer in gewenschte richtingen leiden; de moderne bemestingsleer is onafscheidelijk verbonden aan dezen tak van bacteriologische wetenschap; de zuivering van drink- en afvalwater heeft daarin haar grondslagen te zoeken. Maar ook de talrijke toepassingen, die de menschheid reeds sedert de oudheid van mikrobenwerkingen heeft gemaakt, — de meest primitieve volken kennen de bereiding van alcoholische en zure dranken en winnen vezelstoffen door toepassing van rotingsprocessen — konden eerst bij nadere kennis van die werkingen doelbewust worden verbeterd. Hoe is het beeld van alle gistingsbedrijven sedert de laatste vijftig jaren gewijzigd, welk een omwenteling in de zuivelindustrie, welk een veranderingen in de conservenbereiding, hoe geheel anders staan wij thans tegenover de processen, die verloopen, bij de kuiplooiing, de winning van bastvezels, de tabaks- en de theefermentatie!

Welk een onverwachte gezichtspunten voor de kennis van de levensprocessen in het algemeen heeft deze studierichting ons geschonken. De samenleving van talrijke cellen in een cellenstaat met arbeidsverdeeling en geheel of gedeeltelijk verlies der individualiteit bij de meercellige dieren en planten leidt tot complicaties, die wegvallen bij de eenvoudiger georganiseerde ééncellige wezens, die daartegenover een nog grooter verscheidenheid van vormen en van specialiseering in hun processen te zien geven.

Actief leven bleek mogelijk onder omstandigheden, waarbij men dat vroeger, toen men slechts de hoogere organismen en enkele saprophytische bacteriën tot studieobject bezat, uitgesloten moest achten. Onze denkbeelden omtrent het ontstaan van het leven zijn daardoor geheel gewijzigd en ofschoon wij moeten erkennen nog in het duister te tasten, zoo kunnen onze hypothesen daaromtrent toch op veel hechtere basis worden gegrondvest. De plotselinge veranderingen in uiterlijken vorm en physiologische eigenschappen, die met volkomen zekerheid voor mikroben werden gevonden, openen dan verder uitzicht, om door te dringen in het vraagstuk van de ontwikkeling der organismen van lager tot hooger. Vast staat, dat de mikro-organismen ook in perioden, die meer dan honderd millioen jaar achter ons liggen, een niet minder belangrijke rol speelden dan thans en dat zij het uiterlijk van de aardkorst in sterke mate hebben beïnvloed.

Zoo begint zich de leer der algemeene mikrobiologie te ontwikkelen tot een hecht gebouw, waarvan vele schoone lijnen reeds zichtbaar worden, dat plaats biedt voor toepassingen in techniek, landbouw, hygiëne en huishouding, en van welks steeds hooger rijzende muren zich onverwachte uitzichten openen op terreinen van zusterwetenschappen. Een gebouw, dat reeds nu als een der fraaiste en belangrijkste voortbrengselen van het menschelijk vernuft mag worden aangeduid.

En wanneer wij het geschiedboek van den bouw naslaan, dan treft ons op iedere bladzijde daarvan de naam van den man, tot wiens huldiging wij hierheen zijn gekomen: de naam BEIJERINCK.

Geen onderdeel van dezen jeugdigen tak van wetenschap, waarop niet zijn werken een onuitwischbaren stempel heeft gedrukt. Welk van de talrijke vraagstukken, die ik zooeven in vogelvlucht aan U liet voorbij gaan, men ook nader in oogenschouw neemt, steeds weer blijkt, dat hij den weg heeft gewezen, om het te verbreeden en te verdiepen.

Hoe zou ik in een korte spanne tijds een juist beeld van zulk een werkzaamheid kunnen geven? De eerste verhandeling op bacteriologisch gebied van de hand van BEIJERINCK verscheen reeds in 1887 en zij is door een honderdtal andere gevolgd geworden, waarvan vele een verrassende ontdekking brachten, de meeste geheel nieuwe gezichtspunten openden, alle rijk zijn aan origineele gedachten en treffen door de veelzijdigheid en grondigheid der behandelingswijze. Ik vermag slechts enkele grepen te doen uit dezen rijkdom van materiaal, hopende daarmee toch voldoende BEIJERINCK'S verdiensten voor de wetenschap te belichten.

Wie er mee bekend is, hoe de mikroben bij het bewaren op onze cultuurbodems aan veranderingen onderhevig zijn, zal zich als eersten eisch bij de studie der mikrobiologie stellen: het zoeken naar methoden ter isoleering van bepaalde mikroben uit de natuur. Slechts wanneer men over zulke methoden beschikt, zal men die mikroben in volle activiteit kunnen waarnemen, slechts dan zal men een inzicht kunnen krijgen in hun werkzaamheid onder natuurlijke omstandigheden. Een drietal hoofdmethoden zijn daarvoor in gebruik. De meest toegepaste is wel die, waarbij men, op de door KOCH aangegeven manier, de mikroben, nadat men ze in of over een geschikten cultuurbodem — meest een voedingsvloeistof gestold met gelatine of agar — heeft verdeeld, afzonderlijk tot *koloniën* laat uitgroeien, waarna men dan de koloniën der gewenschte mikroben uitkiest. Maar het bepalen van de *geschiktheid* van den bodem en vooral het uitkiezen van de kolonie biedt vaak onoverkomenlijke bezwaren.

Op hoe vernuftige wijze heeft BEIJERINCK deze isoleeringsmethode verbeterd! Toevoegingen aan de cultuurbodems van geringe hoeveelheden van stoffen, waaruit bepaalde mikroben producten vormen, die door speciale reacties gekenmerkt zijn, maakten het opsporen van deze soorten tusschen talrijke andere, die dat vermogen missen, mogelijk. Welk een fraaie isoleeringsmethode werd bijvoorbeeld verkregen door een toevoeging van *indicaan* aan den gelatinebodem, waardoor onmiddellijk de groep van bacteriën, die uit dit glucoside *indigo* vormen, naar voren kwam, of door toevoeging van querciet, waardoor de aroma-bacteriën (die uit melk aromatische stoffen produceeren) tusschen alle andere dadelijk te herkennen zijn als gevolg van de vorming van pikzwarte velden van geoxydeerd pyrogallol.

Een tweede isoleeringsmethode berust op het vernietigen van sommige bacteriëngroepen en het sparen van andere. Zoo weten wij sedert PASTEUR, dat door kort opkoken van een bacteriën-suspensie alle niet-sporendragende bacteriën worden gedood, terwijl de resistente sporen in leven blijven, die dan, onder gunstige omstandigheden gebracht, kunnen ontkiemen en speciale bacteriën-flora's doen ontwikkelen. Ook hier verrijkte BEIJERINCK de bacteriologie door invoeren van het *lactiseeren*, verhitting van vloeistoffen, die naast andere mikroben, ook melkzuurfermenten bevatten, op circa 65° C., waardoor een speciale groep dier melkzuurfermenten gespaard blijft.

Het grootste succes heeft BEIJERINCK echter ongetwijfeld bereikt door toepassing van de derde isoleeringsmethode door hem als *akkumulatieve* of *electieve* aangeduid, waarbij men, door de ontwikkeling van speciale bacteriën sterk te bevorderen, ééne soort, of een groep van soorten, de overhand doet krijgen boven alle andere, hetgeen dan bij herhaling van het experiment vaak praktisch tot een reincultuur van die bepaalde organismen leidt. Wel is in de werken van PASTEUR een eerste begin van toepassing van zulke ophoopingsproeven te vinden, maar niemand heeft ze zóó consequent ingevoerd als BEIJERINCK. Door die ophoopingsproeven heeft hij de bacteriologie ontdaan van de onzekerheid, die bij toepassing van de beide vorige methoden steeds bleef bestaan: het is daarbij tot zekere hoogte toch een toeval als de gewenschte mikrobensoort wordt gevonden. De ophoopingsproeven daarentegen veroorloven de isoleering van mikroben en het doen optreden van mikrobiologische processen met dezelfde zekerheid, als waarmee de chemicus bij het volgen van bepaalde recepten zijn scheikundige verbindingen ziet ontstaan.

De grootste waarde van deze ophoopingsproeven ligt echter in de omstandigheid, dat zij een inzicht geven in de levenscondities van de naar voren tredende mikrobensoort, terwijl zij doorgaans tegelijkertijd een belangrijk daardoor veroorzaakt proces leeren kennen, een proces, dat ook in de natuur onder den invloed dierzelfde mikroben kan plaats vinden. En bovendien veroorloven zij het qualitatief en zelfs quantitatief vaststellen van de verspreiding der mikrobe. De *oekologie* der mikro-organismen, de leer van hun rol in de natuur, kon slechts door deze ophoopingsproeven tot een studievak worden verheven. De butylalcoholgisting werd door toepassing van dit beginsel gevonden, *Spirillum desulfuricans* als veroorzaker van de zwavelwaterstofvorming uit sulfaten in onze verontreinigde wateren ontdekt, het doen optreden van boterzuurgisting tot een eenvoudig experiment gemaakt, voor de isoleering van azijnbacteriën en melkzuurfermenten konden nu nimmer falende en voor de techniek belangrijke voorschriften worden gegeven, de beteekenis der ureum-splitsende bacteriën werd door hun ophoopingsproef duidelijk, de actieve mikrobe bij de vlasroting werd daarmee gevonden, de bacteriën der denitrificatie — vormers van stikstof en stikstofoxydule uit nitraten en nitrieten — konden alleen daardoor in hun algemeene verspreiding en verscheidenheid worden aangetoond, het voorkomen van de maagsarcine — die tot de grootste en meest opvallende der bacteriën behoort — in den tuingrond kon daarmee worden bewezen.

Merkwaardiger nog werden de met ophoopingsproeven verkregen uitkomsten, toen de samenstelling der cultuurvloeistoffen ingrijpend werd gewijzigd. Nadat de Russische bacterioloog WINOGRADSKY had aangetoond, dat in cultuurvloeistoffen, die geen organische of anorganische stikstofverbindingen bevatten, een sporenvormend organisme in staat is te groeien en vrije atmospherische stikstof tot organische substantie om te zetten, slaagde in het jaar 1901 BEIJERINCK er in, een niet-sporenvormende mikrobe op te hoopen, die datzelfde vermogen bezit en wel zeker in de natuur als stikstofbinder een veel grooter beteekenis bezit. De *Azotobacter chroococcum* — zoo werd deze nieuwe mikrobe genoemd — of daaraan naverwante soorten, zijn sedert, dank BEIJERINCK's hoogst eenvoudige ophoopingsmethode, in alle kultuurbodems op onze aardoppervlakte, waar ter wereld men daar ook naar zocht, gevonden.

Nog tijdens de studie van deze stikstofbindende mikroben, gedeeltelijk met en assistent VAN DELDEN ondernomen, heeft — wederom dank zij de ophoopingsproeven, nu in vloeistoffen, zonder koolstofvoeding — de ontdekking plaats van een mikrobe, die voor zijn koolstofvoeding aangewezen is op de minimale hoeveelheden vluchtige, organische koolstofverbindingen, die in onze atmospheer — vooral waar die door de samenleving van menschen of dieren wordt verontreinigd — voorkomen. Stellig had niemand verwacht, dat de aanpassing van het mikrobenleven zou blijken zóó ver gedifferentieerd te zijn!

Werden uit die ophoopingsvloeistoffen de koolstof- *en* de stikstofverbindingen beide weggelaten en bestonden deze uitsluitend uit een oplossing van enkele stikstofvrije, anorganische zouten, dan trad wel is waar in het donker geen mikrobengroei meer in, maar in het licht ontwikkelde zich een flora van blauwwieren, die als koolstofbron het koolzuur en als stikstofbron de vrije stikstof uit de atmospheer benutten. Een vondst, waarvan de beteekenis nog steeds onvoldoende naar waarde wordt geschat: ziehier toch een mikrobengroei onder de meest elementaire omstandigheden!

Maar laat ik thans een tweede voorbeeld nemen uit de vele onderwerpen, die de bacteriologische werkzaamheid van den jubilaris heeft omvat. Het hangt met het voorafgaande ten nauwste samen.

Bij de mikrobiologische processen, waarbij koolstof-verbindingen worden omgezet, leveren deze omzettingen de bron van energie, waardoor de ontwikkeling der mikroben, de synthese dus van hun lichaamssubstanties, mogelijk wordt. Deze organismen teeren dus op de chemische energie, welke in die organische stof was vastgelegd en welke in laatste instantie blijkt afkomstig te zijn van dat deel van het zonlicht, dat tijdens het proces der koolzuurassimilatie in de groene bladeren werd benut. Alleen bij de laatstgenoemde ophoopingsproef, die voor blauwwieren, werd die zonne-energie rechtstreeks door de mikroben vastgelegd.

Het bleek nu evenwel, dat er mikrobenprocessen mogelijk zijn met geheel andere energie-bronnen, bijvoorbeeld oxydatie van waterstof, zwavelwaterstof, zwavel of thiosulfaten. De mikroben, die dergelijke energiebronnen weten te benutten, ontleenen daaraan dan ook het vermogen, om evenals de hoogere planten, maar nu in het donker, dus zonder zonne-energie, het koolzuur uit de lucht in organische stof om te zetten, teneinde daarmee hun organische lichaamsbestanddeelen op te bouwen.

Hoewel er reeds in oudere en enkele nieuwere onderzoekingen aanwijzigingen voor het bestaan van dergelijke „autotrophe" bacteriën te vinden waren, zijn deze mikrobiologische processen toch eerst in BEIJERINCK's laboratorium, door hemzelf en verschillende zijner leerlingen, onomstootelijk vastgesteld.

Een merkwaardig voorbeeld vormt de denitrificatie van zwavel, nog kort geleden, in 1920, in een belangwekkende verhandeling door BEIJERINCK nauwkeurig beschreven. Is het geen verrassend feit, dat fijngemalen zwavel, verdeeld in een oplossing, waarin alle organische stof ontbreekt, en waarin, naast enkele gewone anorganische zouten en krijt, salpeter als hoofdbestanddeel voorkomt, zich een weelderige mikrobengroei kan ontwikkelen en een bacteriënslijm kan ontstaan, dat zóó rijk is aan organische koolstofverbindingen, dat dit slijm met sterk zwavelzuur verkolingsverschijnselen toont?

Behoef ik u nader te schetsen, dat zulke ontdekkingen onzen gezichtskring buitengewoon verrijken en hoop geven op nieuwe mogelijkheden, die wij thans nog tot het rijk der fabelen brengen?

De verleiding is groot geweest, de beteekenis van nog andere van BEIJERINCK's vondsten op het thans besproken gebied te belichten. Hoe gaarne zou ik u hebben gesproken over zijn waarnemingen omtrent de plotseling en voor immer erfelijk gefixeerde wijzigingen, die mikroben kunnen ondergaan, als ze uit het eene cultuurmedium worden overgebracht in een ander. Deze „physiologische soortvorming" dwingt ons tot grondige wijziging der gangbare opvattingen omtrent erfelijkheid en omtrent de ontwikkeling van hoogere planten of dieren uit de bevruchte eicel. Hoe lokten ook tot nadere behandeling: de belangrijke waarnemingen omtrent mutaties bij mikroben, het plotseling optreden van enkele afwijkende en verder constant zich vermenigvuldigende individuen in een reincultuur, welke mutaties het onderwerp vormden van een der meest uitvoerige publicaties van BEIJERINCK uit de laatste jaren, een verhandeling rijk aan feiten, maar niet minder rijk aan ideeën, die nog in verre jaren zullen vrucht dragen. Hoe gaarne had ik U gesproken over BEIJERINCK's onderzoekingen over wortelknolletjes der vlinderbloemige gewassen, een der eerste onderwerpen, waarmee BEIJERINCK de reeks zijner bacteriologische verhandelingen opende en waarmee hij onmiddellijk zijn naam als bacterioloog vestigde door de isoleering der in de knolletjes voorkomende bacteriën, waarvan hij aantoonde, dat zij als veroorzakers dier aanzwellingen zijn te beschouwen. Een onderwerp, dat hem nimmer heeft losgelaten en waarover hij in 1918 een verhandeling in het licht gaf, waaruit bleek, dat ondanks 30 jaren van onderzoek bij dit probleem nog veel duister is gebleven. Immers ondanks het feit, dat het door de klassieke onderzoekingen van HELLRIEGEL vaststaat, dat de bacteriën der knolletjes van de Papilionaceeën onmisbaar zijn voor de binding der vrije atmospherische stikstof, kon BEIJERINCK zelfs bij gebruik van 1 K.G. der geïsoleerde knolletjes gedurende 12 tot 20 dagen geen spoor van stikstofbinding door zulke knolletjes constateeren. Er moet hier dus aan een zeer indirect, nog geheel onverklaard, verband tusschen het stikstofbindend vermogen van de plant en de aanwezigheid van bacteriënknolletjes gedacht worden.

Maar ik moet thans van zuiver bacteriologische onderzoekingen afstappen, om Beijerinck's overige studiën tot hun recht te laten komen.

Allereerst die op verwant terrein. Geen rijker bron voor het vinden van zoogenaamde *enzymen* dan de mikrobenwereld, maar deze wereld is daarvan geenszins de uitsluitende vindplaats; talrijke enzymen treft men ook in het planten- en dierenrijk aan. Kenmerkend voor de enzymen is, dat zij in hunne werkingen overeenkomst met de levende stof vertoonen, maar daarvan verschillen zij toch in zooverre, dat ze niet, zooals de levende stof, gebonden zijn aan de intacte cel en vooral ook door het uitoefenen van één enkele specifieke functie, in tegenstelling met de levende stof, die toch steeds een complex van werkingen te zien geeft. Vroeger meende men, dat alle enzymen, in tegenstelling met de organismen, in water oplosbaar waren; men sprak wel van „ferments solubles", of kortweg van „fermenten". Men weet thans — en het zijn weer Beijerinck's onderzoekingen, die hier telkenmale licht brachten — dat daarnaast onoplosbare enzymen werkzaam zijn, die suspensies kunnen vormen en dan te vergelijken zijn met de kolloidale metaal-suspensies, waarvan de analogie met de enzymen door Bredig is aangetoond.

De methoden ter bereiding van enzymen zijn door Beijerinck verbeterd; nieuwe proeven, om hun werkzaamheid te demonstreeren, uitgedacht, waaronder die met behulp van lichtbacteriën wel tot de fraaiste experimenten uit de mikrobiologie mogen worden gerekend. Verschillende belangrijke nieuwe enzymen zijn door Beijerinck ontdekt, ik releveer hier alleen de viscosaccharase, een specifiek op rietsuiker werkend enzym; waardoor deze suiker, buiten de bacteriënlichamen, die het enzym voortbrengen, in een voor diffusie niet vatbaar levulaan-slijm wordt veranderd. Ziehier een voorbeeld van een polymeriseerend, dus synthetisch werkend enzym, dat tegenover de veel talrijker, afbouwende enzymen een zeer bijzondere plaats inneemt.

In eene verhandeling, waarvan zich de draagwijdte thans nog niet volledig laat beoordeelen, getiteld „De enzymtheorie der erfelijkheid" en verschenen in 1917, is door Beijerinck de stelling verdedigd, dat de „enzymen" identiek zijn met de „erfeenheden, genen, pangenen of biophoren", welke in de moderne erfelijkheidsleer zulk een alles overheerschende rol spelen en die volgens die leer de dragers der erfelijke eigenschappen in de cel zouden wezen. Deze stoute opvatting opent nieuwe gezichtspunten en zal stellig in komende jaren nog meer aandacht trekken dan ze reeds deed.

Aansluitend aan deze enzymstudiën moeten dan verder de onderzoekingen over verschillende *plantenziekten* genoemd worden. De gomziekte der Prunaceeën heeft reeds van 1884 af Beijerinck's belangstelling gehad en nog in 1914 werd over dit vraagstuk door hem een mededeeling aan de Kon. Akademie te Amsterdam gedaan. Een wondprikkel, die op velerlei wijzen kan worden veroorzaakt of versterkt, o.a. door de werking van *Coryneum Beijerinckii*, maar die ook volkomen normaal kan wezen, bleek oorzaak voor de vergomming van bepaalde weefsels in Pruim, Kers, Abrikoos en verwante gewassen.

Baanbrekend waren Beijerinck's studiën over de besmettelijke mozaikziekte der tabaksplanten, welke dateeren uit het jaar 1898. Ondanks nauwgezet onderzoek kon geen mikrobe als verwekker daarvan worden aangewezen en toch lieten zich de kenmerkende vlekken op de bladeren door inenting met ziek materiaal op gezonde planten te voorschijn roepen. De smetstof was door de fijnste bougies, die alle bacteriën tegenhielden, zelfs door agar, filtreerbaar; zij was echter op geen voedingsbodem te cultiveeren. Hier aarzelde Beijerinck, toen zijn overtuiging vast stond, niet om te spreken van een „contagium fluïdum", wel te onderscheiden van een enzym, want van de smetstof moest worden aangenomen, dat zij niet slechts katalytisch, door haar aanwezigheid alleen, werkt, maar ook, dat zij zich in het plantenlichaam vermeerdert. Heftig is Beijerinck over deze opvatting aangevallen; velen zwegen, maar twijfelden. Beteekende die opvatting toch een teruggrijpen naar een denkbeeld, waarmee men na Pasteur's onderzoekingen voor immer meende te hebben afgerekend. Maar sedert zijn de gevallen, waarin dergelijke vloeibare contagiën als werkzame agentiën moeten worden aangenomen, steeds talrijker geworden. De oorzaak van het mond- en klauwzeer, de smetstof van pokken en roodvonk, van mazelen en gele koorts, bleken even zoovele voorbeelden daarvan. Het moet voor Beijerinck een groote voldoening zijn, de erkenning van zijn denkbeeld te constateeren.

Na Beijerinck als mikrobioloog — in den ruimsten zin — Beijerinck als botanicus. Men meene niet, dat ik de botanische werken van den jubilaris bij zijn bacteriologische ten achter stel. Zij bieden den deskundigen lezer geen minder groot genot, maar ze spreken door den specialen aard niet zoo sterk voor den buitenstaander. Had ik een chronologische volgorde bij mijn overzicht in acht genomen, dan zou ik de botanische studiën voorop hebben moeten stellen. Ik had dan moeten schetsen, hoe Beijerinck reeds in de jaren, waarin hij de H.B.S. met 5-jarigen cursus te Haarlem bezocht, d.i. van 1864 tot 1868, een groote liefde voor natuurkennis aan den dag legde. Hoe hij toenmaals, in 1867, den eersten prijs behaalde bij de beantwoording van een prijsvraag, uitgeschreven door den heer KRELAGE te Haarlem, waarbij gevraagd werd: de inzending van 150 gedroogde planten uit de omgeving dier stad met vindplaats, datum, Latijnschen en Nederlandschen naam. Was het wonder, dat de prijs: een zilveren medaille en de bekende „Flora van Nederland" van OUDEMANS, den jeugdigen florist tot spoorslag waren, om zijn plantkundige studiën voort te zetten? Ondanks alle andere onderzoekingen is hij de botanie ook in zijn verder leven nimmer ontrouw geworden.

Wel werd na het behalen van het eindexamen de studie voor technoloog ter hand genomen en daarbij de grondslag gelegd voor de chemische kennis, die den lateren mikrobioloog zoo zeer te stade zou komen. Wel werden met zijn vriend VAN 'T HOFF chemische proeven op de studentenkamer genomen: o.a. ossengal gekookt met zoutzuur op het potkacheltje, wat het vernielen van behang en meubilair meebracht. Maar daarnaast bleef tijd over tot natuurhistorische studiën en werden skeletten, door afkoken van dieren, vervaardigd en botanische excursies naar Rijswijk en verdere omstreken van Delft ondernomen.

Ik mag als bekend onderstellen, hoe BEIJERINCK, na het behalen van den titel van technoloog, gebruik makend van het door THORBECKE uitgelokt Besluit, zonder staatsexamen toelating verkreeg tot de Universiteit en hoe hij in 1877 te Leiden den graad van Doctor in de Wis- en Natuurkunde behaalde. Proefschrift en eerste publicaties zijn van botanischen aard, of juister, betreffen een grensgebied tusschen botanie en zoölogie: dat der „galvormingen". Wie eenig modern werk over de vliesvleugelige insecten opslaat, zal daarin zeker verschillende, van een meesterhand getuigende, teekeningen, met tot aanduiding als auteur: BEIJERINCK, aantreffen.

Sta mij toe eenige der vele problemen, waarover het hier handelt, toe te lichten. Een ieder kent de bekende gallen aan de onderzijde der gewone eikenbladeren en ieder weet ook, dat ze een larve van een galwesp bergen. Maar minder bekend is het, dat als die galwesp laat in het najaar uit de gallen, die dan op de afgevallen bladeren te zoeken zijn, te voorschijn komt, zij uitsluitend vrouwelijke individuen te zien geeft, welke zonder bevrucht te zijn, parthenogenetisch dus, eieren voortbrengen. Die eieren worden afgezet in slapende knoppen aan de basis van jonge eikenboompjes. In het volgende voorjaar ontwikkelt zich dan uit dien knop een kleine gal, niet langer dan een 0.5 cM., violet van oppervlak, die in Mei en Juni tot rijpheid komt. Uit zulke galletjes kruipen dan kleine, zwarte galwespen; ditmaal echter mannetjes en vrouwtjes, die tamelijk belangrijk van de parthenogenetische insekten afwijken en dan ook aanvankelijk voor een andere soort werden gehouden. Al spoedig volgt de paring en de vrouwtjes zetten de eieren af in de nerven aan de onderzijde van nog jeugdige eikenbladeren, waaruit zich dan in den loop van den zomer de U bekende gallen ontwikkelen.

Deze regelmatige afwisseling van een generatie van enkel wijfjes met eene van mannelijke en vrouwelijke dieren is voor tal van typische galwespen geconstateerd. Een zeer merkwaardige ontdekking was het nu — wij danken haar met vele andere over de biologie der gallen en der galwespen al weder aan BEIJERINCK — dat er galvormingen bestaan, waarbij de twee generaties der galdieren niet één enkele plantensoort — zooals zooeven, waar zich alles aan den gewonen eik afspeelde — maar constant twee verschillende plantensoorten voor hun gallen noodig hebben. Met generatiewisseling gaat dan waardwisseling gepaard. De merkwaardige *knoppern* zijn uitwassen op de nap der gewone eikels. Zij zijn het gevolg van de afzetting van een ei door een bevrucht vrouwelijk insekt, dat de legboor steekt door het napje van den jongen eikel van onzen meest gewonen eik, *Quercus pedunculata*. Binnen de zich dan vormende knopperngal ontwikkelt zich een parthenogetisch insekt, dat in begin Maart van het volgend jaar uitkruipt, en de eieren afzet in de nog geheel gesloten knoppen van mannelijke eikenbloempjes, niet echter van den gewonen eik, maar van den mos-eik, *Quercus Cerris*. Wanneer men daarbij bedenkt, dat die mos-eik, die in Zuid-Europa thuis hoort, hier te lande slechts op zeer enkele plaatsen voorkomt en dat het uitkruipen en eierenleggen zich in het tijdsverloop van enkele dagen afspeelt, dan krijgt men een denkbeeld, hoe moeilijk de ontwarring van deze zeer gecompliceerde verhoudingen was.

De generatiewisseling — trouwens volstrekt niet aan alle galdieren eigen — is slechts één der tallooze problemen, die de galstudie oplevert. Zij biedt nog vele andere, zooals: de groote verscheidenheid der vormen en de niet minder groote verscheidenheid in dierlijke en plantaardige verwekkers, de anatomische bouw, de parasietische medebewoners, de chemische samenstelling, de technische toepassingen en vooral de merkwaardige aanpassingsverschijnselen, zoowel bij het galinsekt als bij de galvorming zelve. Zien wij niet hier een geval, waarin de plant voor het ten hare kosten levende insekt vaak beschuttende cellagen vormt, soms een speciaal, voedselrijk weefsel, dat dienst doet bij het grootbrengen van het dier, hetwelk haar naderhand opnieuw met gasten zal bevolken? Hoe ook wel door de plant voor wateraanvoer of luchttoetreding, in de gal, soms zelfs voor het op tijd openen van die gal tot vrij-laten van het insekt, en voor nog veel meer, wordt zorggedragen? Ziehier een gebied, rijk aan raadselen, telkens aanleiding gevend tot het belichten van algemeen-biologische problemen. Een gebied waardig aan een BEIJERINCK, die ook door zijn galstudiën als een onzer beste natuuronderzoekers mag genoemd worden.

Geenszins zijn die studiën het eenige, wat BEIJERINCK op botanisch gebied verricht heeft, maar ik mag thans niet langer voor wetenschappelijk werk Uw aandacht vragen. Een kort woord nog over BEIJERINCK als docent en als leermeester.

Al is het thans reeds 36 jaren geleden, sedert BEIJERINCK als docent voor de botanie aan de Wageningsche Hoogere Landbouwschool werkzaam was, men ontmoet nog vaak personen, die met groot enthousiasme over zijn onderwijs aldaar spreken. Wie het voorrecht heeft gehad, zijn colleges te volgen aan de Polytechnische School, waaraan hij sedert 1895 vast is verbonden, — na daar reeds vroeger als privaat-docent voordrachten van plantkundigen aard, met de nijverheid in verband staande, te hebben gegeven — hij zal met niet minder warmte dat onderwijs roemen.

Toegelicht met schitterende experimenten, uitmuntende door eenvoud en overzichtelijkheid, in een heldere voordracht, waarbij ieder woord van de groote liefde voor het gedoceerde vak sprak, verklaard met nieuw geteekende, smaakvolle collegeplaten, werden de belangrijkste onderwerpen van het nog steeds zich ontwikkelend leervak besproken. Welk een bekoring ging voor de jeugdige toehoorders van zulk een schittering van geest uit, hoe wist BEIJERINCK de liefde voor de levende natuur bij ons te wekken en hoe verstond hij het ons de lastigste problemen te verduidelijken. Gij, BEIJERINCK, had U trouwens een bijzonder dialekt eigen gemaakt, om ons toch maar goed bij te brengen, dat de zaken, waarover Gij sprak, zulke waren, die wij dagelijks om ons heen zagen en van den meest huishoudelijken aard. Het heugt mij nog als de dag van gisteren,al is het 22 jaar geleden, hoe Gij op het eerste college, dat ik van U bijwoonde, Uw toehoorders aanraadde, te nemen: een likkie flap uit de sloot en dat te smeren in een kolfie en hoe Gij hun dan voorspelde dat, wat zij zouden zien was: nies, heelemaal nies.

Wie, zooals ik, het geluk had, vele jaren Uw assistent te zijn en wien een blik gegund is geweest in de geheimen van de heksenkeuken,waarin de bereiding der terecht met zooveel graagte door de wetenschappelijke wereld ontvangen spijzen plaats vond, hij moet zich bedwingen, om niet in beschouwingen te treden over de wijze, waarop Uwe ontdekkingen tot stand kwamen, beschouwingen, die de wisselwerking van geniale invallen, onvermoeiden speurzin, scherp combinatievermogen, hartstochtelijken aandrang voor het oplossen van eenmaal gestelde problemen, zouden moeten omvatten.

Maar laat ik er mij voor wachten, mijn toehoorders nogmaals met wetenschappelijke vraagstukken te vermoeien. Er zouden uit die jaren, waarin de aandacht der wetenschappelijke wereld buiten onze grenzen steeds meer op BEIJERINCK's werken werd gevestigd, waarin van heinde en verre — uit Moskou, uit Ascension in Paraguay, uit Bergen in Noorwegen, uit Tokyo in Japan — de leergierigen naar Delft reisden, om zich onder Uw leiding te stellen, nog heel wat herinneringen zijn op te halen. Vaak ook van humoristischen aard. Want zulke vogels van diverse pluimage voegden zich wel eens lastig naar Uw strenge regels bij het dagelijksche werk. Ligt er geen humor verscholen in Uw gebod aan een Weenschen hoogleeraar, die met een witte laboratoriumjas aankwam, om die „Schmierjacke" uit te laten, waarna dan de opdracht aan den assistent werd gegeven, om dien hoogleeraar nu ook te zeggen, verder zijn kraakschoenen thuis te laten?

Uw assistent werd Uw collega, het contact losser, de samenkomsten zeldzamer. Maar wanneer zich onze wegen kruisten, bleek Uw warme vriendschap, die ik met groote dankbaarheid heb te gedenken. Hoe vonden trouwens allen, die Uw belangstelling genoten, in U een bron van verfrissching voor hun geest, hoe genoten zij van Uw veelzijdig, oorspronkelijk intellect. Al ontvloodt Gij doorgaans het gezelschappelijke leven, Uw komst daar was een feest. Hoe wist gij dan allen door het verdedigen van paradoxale stellingen in Uw ban te voeren.

Hooggeschatte jubilaris! Uw feestredenaar heeft zijn taak volbracht. Hij is er zich van bewust, dat hij de velerlei facetten van Uw geest slechts ontoereikend deed spiegelen, toen hij U schetste als bacterioloog, enzymoloog, phytopatoloog, botanicus, zoöloog, docent, leermeester en vriend. Aan een meer deskundige is het toevertrouwd, weldra nog een ander kristalvlak van dezen edelsteen te doen schitteren.

Laat mij U thans als slotwoord nog verzekeren, hoe noode Uw Delftsche vrienden en vereerders U van hier zien vertrekken. Zij troosten zich met de overtuiging, dat Gij nog steeds met onverflauwde toewijding Uw studiën vervolgt en dat de wetenschap U mag behouden. In een natuur, die rijker gezegend is dan de Delftsche omgeving, zullen nieuwe vragen Uw aandacht boeien. Wij zullen ons gelukkig achten, als Gij ons toestaat, ons zoo nu en dan te warmen aan het jeugdig vuur, dat nog immer in U brandt.

Appendix H.

Abstract from the lecture given by BEIJERINCK on the occasion of his retirement from the chair at the "Technische Hoogeschool" on May 28th, 1921. *)

Na een korte inleiding zegt spreker: Bij vergelijking van de hoogere wezens met de lagere, blijkt, dat al wat leeft zes elementaire functies gemeen heeft: Voeding, ademhaling of gisting, groei, celdeeling, erfelijkheid en variabiliteit. Prof. BEIJERINCK wenschte in dit uur slechts te spreken over de Erfelijkheid en de Variabiliteit bij de microben. Dit zijn de meest ingewikkelde levensprocessen, maar bij de microben komen zij in den meest eenvoudigen vorm voor.

In 1846 was door HUGO VON MOHL het protoplasma als het eigenlijke levende deel der cel herkend. De door ROBERT BROWN in 1833 ontdekte celkern bleek daarvan een integreerend deel te zijn en de verdere studie van protoplasma en kern heeft tot de protoplasmatheorie van het leven gevoerd, welke tegenwoordig de grondslag is van verschillende biologische wetenschappen, zooals de physiologie, de anatomie, de microbiologie, de embryologie en de pathologie. Dit wil spreker thans nader uiteen trachten te zetten en aanwijzen, hoe de Erfelijkheid en de Variabiliteit daarmede in verband staan. Het zal noodig zijn, daarvoor de cel te beschouwen in de verschillende graden van volkomenheid, welke daarvan gevonden worden eenerzijds bij de hoogere wezens, anderzijds bij de microben.

Daarna bespreekt prof. BEIJERINCK het erfelijksheidsvraagstuk bij de hoogere wezens en licht dit toe aan de onderzoekingen van MORGAN, neergelegd in zijn werk: „A Critique of the Theory of Evolution", Princeton, London, Oxford 1916 (een diamant onder het vele kiezel op de boekenmarkt). Hierin wordt aangetoond, hoe bij de vlieg *Drosophila ampelophila*, alle kenmerken vertegenwoordigd worden door evenzoovele chromomeren, welke tezamen de 4 chromosomenparen van de celkern vormen. MORGAN neemt aan, dat er in het geheel omstreeks 2500 chromomeren in de chromosomen voorkomen, zoodat het insect ook zooveel zelfstandige eigenschappen heeft. Over de natuur der genen of MENDELsche factoren handelt het onderzoek van MORGAN echter niet. Daarvoor, zegt spreker, is ook juist het arbeidsveld van het Laboratorium voor Mikrobiologie, en hij verdedigt de stelling, dat de factoren of genen van de erfelijksheidsonderzoekers niets anders dan enzymen zijn. Elke zelfstandige eigenschap beantwoordt dientengevolge aan een specifiek enzym en aan een specifieke chromomere. Het ligt voor de hand, om te zeggen, dat deze chromomere het betrokken enzym voortbrengt. Spreker noemt alle chromomeren tezamen het genoplasma en alle genen, factoren of enzymen tezamen het phaenoplasma, waarbij hij zich aansluit aan de beteekenis van de woorden phaenotypus of genotypus van JOHANSSEN.

Overgaande tot de behandeling van de microbencel, wordt gewezen op den eenvoud daarvan en op het merkwaardige feit, dat vele lagere microben alleen uit genoplasma schijnen te bestaan. Dit kan echter niet zoo zijn, want juist door hun rijkdom aan enzymen bewijzen deze microben ook veel phaenoplasma te bezitten.

De verschillende vormen der microbencellen besprekende, wordt aangetoond, dat de differentieering daarvan bij de phylogenetische ontwikkeling heeft plaats gehad van den eenvoudigsten *Micrococcus* uitgaand, ten eerste doordat nieuwe richtingen van celdeeling zich bij de oude voegden, waarbij *Sarcina* ontstond: ten tweede doordat ciliën of bewegingsorganen werden verkregen: ten derde, door inwendige sporenvorming.

Gewezen wordt op den eenvoud van het vraagstuk der Erfelijkheid bij de microben, vergeleken met dat der hoogere wezens: bij de laagste vormen valt het geheel samen met de celdeeling, waarbij twee dochtercellen ontstaan, welke uit den aard der zaak identiek met de moedercel kunnen zijn.

Het vraagstuk der veranderlijkheid bij de microbe is daarentegen veelzijdig. De

---

*) Reprinted from the "Nieuwe Rotterdamsche Courant" of May 28th, 1921 (Ev.Ed.).

theorie van het phaeno- en van het genoplasma geeft aanleiding veranderingen in het eerste en de fluctueerende variabiliteit van het laatste met de constante variabiliteit of de mutatie van HUGO DE VRIES in verband te brengen. De laatste kan echter op twee totaal verschillende oorzaken berusten, namelijk, òf op een heterotype celdeeling, waarbij de dochtercel niet precies hetzelfde genoplasma ontvangt als de moedercel, òf op een blijvende verandering van genoplasma door uitwendige invloeden. Bij vele microben geeft bijv. kultuur boven het optimum der groeifunctie aanleiding tot diep-ingrijpende erfelijk standvastige verandering.

Beschreven werd een nieuw ontdekt enzym, dat door sommige azijnbacteriën en melkzuurfermenten wordt afgescheiden en het vermogen bezit van het reeds door dr. BARENDRECHT opgemerkte verschijnsel van het uitvlokken van gist te veroorzaken. Het genoplasma van dit enzym, dat verder flokkase zal genoemd worden, is uiterst gevoelig voor warmte. Door cultuur der betrokken bacteriën bij 40° gaat het vermogen om het enzym te vormen, volkomen verloren.

Maar de wetenschap is lang en onze tijd is kort — aldus besluit spreker. Ik roep een vaarwel toe aan onze studenten, met wie ik zoo gaarne samenwerkte, aan alle collega's in het bijzonder aan die mijner afdeeling. Aan de Technische Hoogeschool die mij, door hare rijke hulpmiddelen en door een verlicht en welwillend curatorium, waarbij mijn gedachten in de eerste plaats teruggaan tot de heeren CLUYSENAER en DE VOGEL, mij een aangenaam professoraal leven heeft gegeven. Want ik heb mij, zoover als mijn faculteiten dit toelieten, kunnen verdiepen in de schoonheid mijner wetenschap; en U, mijne heeren- en dames-studenten, kan ik de verzekering geven, dat de wetenschap voor ieder, die zich daaraan met toewijding geven kan, de hoogere poëzie van het leven is. Ook roep ik een vaarwel toe aan mijn laboratorium, en hoop, dat het, na mijn vertrek, een nieuw tijdperk van bloei tegemoet gaat.

Als het blad van den boom valt, aldus besluit prof. BEIJERINCK, geschiedt dit doordat zich een scheidingslaag van levend celweefsel heeft gevormd tusschen tak en blad.

Bij het afvallen spouwt de scheidingslaag zich in tweeën, waarbij een drukwerking de vaatbundels, dat zijn de verbindingsdraden tusschen tak en blad, verbreekt. De eene helft der scheidingslaag blijft aan den tak, de andere aan het blad.

De boom is de T. H. en de tak is de afdeeling, het afvallende blad is de vertrekkende hoogleeraar, de druk die de scheiding veroorzaakt is de wet.

De verdubbelde scheidingslaag is de herinnering. Deze zal beiderzijds eenigen tijd blijven voortbestaan, op den tak, in de afdeeling, tot den groei daarvan de laatste sporen zal hebben uitgewischt. Dit zal lang duren, want in de gedenkboeken der T. H. van later tijd zullen zij, die nà ons komen, onze namen vinden en zich afvragen: wie waren zij?

Maar het blad met zijn scheidingslaag vergaat spoedig, gelijk de scheidende hoogleeraar, die de herinnering medeneemt tot het oogenblik, dat hij zelf zal ophouden te zijn.

Appendix I.

Speeches held by Professor G. VAN ITERSON JR. and by Professor A. J. KLUYVER on June 14th, 1927, on the occasion of the golden jubilee of BEIJERINCK's doctorate. *)

De Voorzitter van het Comité, Prof. Dr. Ir. G. VAN ITERSON JR., opende de bijeenkomst met de volgende toespraak:

Mijnheer de Voorzitter van het College van Curatoren der Technische Hoogeschool,

Dames en Heeren.

Het zal wel niet te veel gezegd zijn, wanneer ik constateer, dat wij allen, zonder onderscheid, die hier aanwezig zijn, trotsch zijn op de beteekenis, die onze Hoogeschool in den loop der jaren voor ons vaderland heeft gehad en die zij ook heden ten dage daarvoor bezit. De ontwikkeling van de industrie en der toegepaste natuurwetenschap hier te lande en ook in Nederlandsch-Indië is nauw verbonden aan de werkzaamheid van de ingenieurs, die van onze instelling de grondslagen meekregen, waarop zij voortbouwden en die hen in staat stelden, bij te dragen tot de welvaart van het land.

Dit heugelijk resultaat is stellig niet in de laatste plaats te danken aan de werkzaamheid van vele verdienstelijke docenten, die met liefde voor hun vak en met opoffering van persoonlijke neigingen en genoegens onvermoeid streefden naar volmaking van het onderwijs aan deze Hoogeschool.

Naarmate men ouder wordt en de grenzen van zijn kennis gaat afbakenen en den oorsprong daarvan onbevooroordeeld vastleggen, neemt in den regel ook de erkentelijkheid toe voor de leermeesters, die uit den schat van hun weten rijke gaven uitdeelden, die den leergierigen jongeling bevrediging schonken, die hem den weg wezen naar dat zeer bijzondere geluk, dat men alleen door zelfstandig denken en eigen onderzoek kan ondervinden. Wie onzer zal niet onmiddellijk de namen kunnen noemen van de enkele personen, die nog steeds hun invloed op ons denken doen gelden en aan wie wij ons voor altijd verplicht gevoelen voor het hoogste geestelijk genot, dat wij ondervonden?

In dezen zin zullen vandaag de gedachten van velen hier te lande en in verre gewesten uitgaan naar den man, die op zijn uitdrukkelijk verlangen in de stille rust van de schoone natuur, waarheen hij zich terugtrok, kan terugzien op 50 jaren noesten arbeid.

Van de ver strekkende beteekenis van het werk van MARTINUS WILLEM BEIJERINCK voor wetenschap en techniek heb ik eenige jaren geleden een beeld trachten te ontwerpen, waarbij ik mij bewust was, dat het slechts op onvolkomen wijze recht deed wedervaren aan de verdiensten van een man, dien het nageslacht zal erkennen als een baanbreker en een wegwijzer voor tal van latere onderzoekers. Uit den merkwaardigen levensloop van BEIJERINCK, gedurende de halve eeuw, die verloopen is sedert het verschijnen van het uitnemende Proefschrift „Bijdrage tot de Morphologie der Plantegallen" op 14 Juni 1877, zullen aanstonds door BEIJERINCK's opvolger enkele bijzonderheden worden meegedeeld, waardoor BEIJERINCK's werk nog nader belicht zal worden en stellig is niemand beter dan collega KLUYVER in staat om de onschatbare waarde van dat werk te schetsen.

Mij zij het vergund, hier met een enkel woord de herinnering op te roepen aan den *persoon*, die op den levensgang van velen onzer een grooter invloed heeft uitgeoefend dan hij zich zelf wel bewust is.

Het was een gelukkig denkbeeld van onzen collega, Professor A. W. M. ODÉ, toen hij BEIJERINCK kort vóór zijn vertrek uit Delft verzocht, te willen poseeren voor een plaquette. Wij vrienden en vereerders van BEIJERINCK kunnen den kunstenaar niet

*) Reprinted from Chem. Weekbl. **24**, 330, 1927.

dankbaar genoeg zijn voor zijn initiatief en voor de uitnemende wijze, waarop hij zijn werk heeft ten uitvoer gebracht. De reproductie op kleine schaal, die hier aanwezig is, geeft uiteraard slechts een onvolkomen indruk van de plaquette, die thans is bevestigd in het voorportaal van het Laboratorium voor Microbiologie dezer Hoogeschool, dat in 1897 naar de aanwijzingen van BEIJERINCK is gebouwd en waarin zoo menige schitterende ontdekking op bacteriologisch en algemeen biologisch terrein werd gedaan. Voor de geschiedenis der bacteriologie en der wetenschap in het algemeen is dit gebouw een gewijde plek en de beeltenis van den man, aan wien wij dit danken, behoort daar geplaatst en in die omgeving gezien te worden.

Wie Uwer aanstonds ter plaatse deze uitstekend getroffen beeltenis aanschouwt — een beeltenis, die ook de volle waardeering van den jubilaris van heden verwierf — zal niet getroffen worden door de hooge intelligentie, het scherpe waarnemingsvermogen, de rustige beschouwingswijze en den kritischen zin, die uit dezen expressieven kop tot ons spreken? Voorwaar, een schitterend pendant voor de beeltenis, die op geen honderd meter van hier tegen het tuinhek van het aangrenzend gebouw is aangebracht, van den, ook door BEIJERINCK zoo hoog vereerden, voorganger: van ANTONI VAN LEEUWENHOEK! Wel is het een dier merkwaardige spelingen van het toeval te noemen, waarvan de geschiedenis verschillende voorbeelden kent, dat twee coryphееën op hetzelfde terrein der natuurwetenschappen, met een tusschenruimte van 300 jaren, in hetzelfde stadje hun groote vondsten deden — zij 't ook, dat hun werkwijze, hun geaardheid en hun levensgang in vele opzichten uiteenloopen.

En welke herinneringen roept dit beeld van BEIJERINCK niet op bij hen, die hem persoonlijk kennen en onder zijn zoo sterk-stimuleerenden invloed de biologie leerden beoefenen? Wie van hen denkt niet bij het zien van deze geestrijke beeltenis aan den altijd springenden bron van vernuft, aan den onvermoeiden uitlegger van gecompliceerde verschijnselen en aan den onovertroffen vertolker van de denkbeelden, die de contemplatie der natuur bij hem opwekte? Wie van hen voelt niet het verlangen terugkeeren naar den sterken prikkel, die van zijn kritiek uitging, naar den caleidoskoop van verrassende gezichtspunten, die BEIJERINCK door zijn zoo veel omvattende kennis der natuur ons wist voor te tooveren? Wie verlangt niet terug naar het medemaken van de koene gedachtesprongen, die de voor indrukken zoo vatbare geest tentoonspreidde, want lag niet voor ons leerlingen een groote bekoring in de emotionaliteit van dit helder en logisch verstand en is daarin ook niet een der oorzaken aan te wijzen, waaruit wij BEIJERINCK's liefde voor de natuur en zijn fraaiste wetenschappelijke vondsten hebben te verklaren?

Zoo zullen wij ouderen dit beeld bezien met de piëteit van dankbare leerlingen voor een beminden leermeester of met de warme gevoelens, die men gevoelt voor een mensch, die men met trots tot zijn vrienden rekent.

Voor de jongeren, die BEIJERINCK niet persoonlijk kenden, moge het feit, dat wij vereerders en vrienden van BEIJERINCK ons geroepen voelden, den dag van heden voor het nageslacht vast te leggen, een aansporing zijn tot kennisnemen van zijn werk en tot navolging van zijn voorbeeld. En zoo zou dan ook het woord dat van VOLTAIRE werd gesproken, als onderschrift passen voor het beeld van den man, dien wij eeren:

„Qui que tu sois, voici ton maître,
Il l'est, le devient ou le doit être".

Mijnheer de Voorzitter van het College van Curatoren.

Het Comité, dat zich in 1921 tot huldiging van BEIJERINCK op zijn zeventigsten verjaardag vormde, beschouwde het als een onmisbare aanvulling van de toenmaals met vreugde volbrachte taak, om heden het gouden doctorsfeest van onzen vereerden leermeester te herdenken. Dat Comité werd destijds gepresideerd door een der oudste vrienden van den jubilaris, Prof. HOOGEWERFF, die mij met het oog op zijn leeftijd verzocht heeft, thans zijn functie waar te nemen. Sta mij toe U, uit naam van het Comité, oprechten dank uit te spreken voor Uw bereidheid om de plaquette te aanvaarden en haar te doen aanbrengen op de plaats, die haar toekomt. Wij hebben er trouwens niet aan getwijfeld, of Uw College, dat in zoo hooge mate deelt in het wel en wee onzer Hoogeschool, zou ook in deze met ons medeleven.

De reproductie op verkleinde schaal, die hier is opgesteld, moge Uw College een plaats geven tusschen de beeltenissen van curatoren en docenten op het Hoofdgebouw onzer Hoogeschool. Aan hen, die het Laboratorium voor Microbiologie aan de Nieuwe Laan slechts zelden betreden, zal het welkom zijn, ook elders de beeltenis te vinden van den vereerden docent, die er zooveel toe heeft bijgedragen, dat de naam onzer Hoogeschool met roem over de wereld werd verspreid.

Ik mag hier dan voorts vermelden, hoe een tweede exemplaar van deze reproductie bestemd is geworden voor 't Microbiologisch Laboratorium der Landbouw-Hoogeschool te Wageningen. Ons Comité meende daarmee niet alleen de herinnering te moeten levendig houden aan de jaren, waarin BEIJERINCK aan de toenmalige Hoogere

Landbouwschool zijn, ook daar zoo hoog geprezen, onderwijs gaf, het Comité wilde tevens doen uitkomen, hoe het genoemde Laboratorium te Wageningen door nauwe banden verbonden is aan het Delftsche. Is niet de fraaie, nieuwe instelling te Wageningen gebouwd en wordt ze niet geleid door een der bekwaamste leerlingen en warmste persoonlijke vrienden van BEIJERINCK, den eersten promovendus van den jubilaris, den eersten promovendus ook van onze Hoogeschool, door Prof. SÖHNGEN?

Veroorloof mij, hieraan een woord van erkentelijkheid toe te voegen, gericht tot hen, die hierheen zijn gekomen. Het heeft ons moeite gekost, den jubilaris te bewegen, ons toe te staan, dezen dag niet geheel ongemerkt te laten voorbijgaan en wij hebben zijn toestemming slechts gekregen, toen wij hem toezegden, dat deze plechtigheid een intiem karakter zou dragen en de uitnoodigingen tot kleinen kring beperkt. Hadden wij aan de gebeurtenis die wij herdenken, de ruchtbaarheid gegeven, die zij ongetwijfeld verdient, dan zouden nog velen met ons hier aanwezig zijn. Wij hebben noode buitenlandsche deputaties van een komst hierheen weerhouden.

Maar het verheugt ons Comité toch in hooge mate, hier vele personen te zien, die aan BEIJERINCK na staan en wier tegenwoordigheid hier een bewijs is, hoe de herinnering aan hem, ook al is de mogelijkheid om hem te ontmoeten, moeilijker geworden, levendig is gebleven. Wij zijn er zeker van, dat het den jubilaris aangenaam zal aandoen, wanneer wij hem aanstonds het album met handteekeningen van vrienden en vereerders, die ons hielpen het plan voor heden te verwezenlijken, zullen overreiken, een album, dat door de kunstzinnige hand van Mejuffrouw J. B. MOUTON, de bekwame assistente voor decoratieve kunst aan onze Hoogeschool, op zulk een artistieke wijze is versierd en van opdracht voorzien. Wie de voorliefde kent, waarmee BEIJERINCK gedurende de afgeloopen 50 jaren telkenmale tot de studie der galvormingen terugkeerde, zal begrijpen, hoe de teekening op den band niet slechts het onderwerp van het Proefschrift verzinnebeeldt, maar ook de aanduiding is van een der moeilijkste en rijkste problemen, die den grooten bioloog tot den huidigen dag in den ban hielden.

De verleiding is groot om te spreken over sterke banden, die BEIJERINCK bonden en binden aan personen, wier namen in dat album voorkomen. Ik wil daaraan weerstand bieden en hier alleen een woord spreken tot de vertegenwoordigers van de Universiteit te Leiden om hun te verzekeren, hoezeer wij hun overkomst op prijs stellen. Zij mogen het ons ten goede houden, dat wij den ,,Leidschen" doctor in zulk een mate voor onze Hoogeschool hebben opgeëischt. Ligt echter niet daarin een erkenning van het vele, dat de Leidsche Universiteit tot de vorming van den jeugdigen geleerde bijdroeg? Misschien zou er meer aanleiding zijn, onze verontschuldiging aan te bieden voor het feit, dat de Delftsche School Leiden beroofde van een uitstekend candidaat voor den door het overlijden van Prof. SURINGAR vacant gekomen leerstoel voor botanie. Maar de wijze, waarop BEIJERINCK hier in staat is gesteld, zijn groote gaven te ontplooien en de vereering, die wij hem hier toedragen, zal het der Leidsche Universiteit gemakkelijker maken, onze toenmalige stoutmoedigheid te vergeven en bij haar de overtuiging vestigen, dat de promovendus, aan wien zij 50 jaren geleden den doctorsbul met zulke groote verwachtingen uitreikte, hier de plaats heeft gevonden, den eminenten leerling der Leidsche Universiteit waardig!

Ik twijfel niet, of ik zal de tolk zijn van Uw aller gevoelens, wanneer ik aanstonds den jubilaris uit Uw naam toewensch, dat hij nog lang getuige moge zijn van onze onvergankelijke vereering en vriendschap.

Vervolgens gaf de Voorzitter van het College van Curatoren, Prof. Dr. Ir. J. KRAUS, uiting aan de gevoelens van groote erkentelijkheid van dit College voor het kostbare en fraaie geschenk, waarbij hij nogmaals de verzekering gaf, dat het werk van BEIJERINCK aan de Technische Hoogeschool ook door Curatoren hoogelijk is gewaardeerd en bewonderd.

Hierna hield Prof. Dr. Ir. A. J. KLUYVER de hieronderVolgende toespraak:

Mijnheer de Voorzitter van het College van Curatoren,

Dames en Heeren.

Wanneer ook ik nog een oogenblik Uw aandacht verzoek, dan geschiedt dit in de eerste plaats, omdat het passend lijkt nog even te verwijlen bij de keten van gebeurtenissen, die den 14den Juni 1877 verbindt met ons huidig samenzijn. Het schijnt toch op het eerste gezicht bevreemdend, dat een promotie aan de Leidsche Universiteit, na verdediging van een proefschrift, dat tot sobere titel droeg: ,,Bijdrage tot de morphologie der plantegallen", vijftig jaren later leidt tot een plechtige bijeenkomst in de vergaderzaal eener Technische Hoogeschool.

Maar ook hij, die zich nader in de geschiedenis van dit gebeuren verdiept, ontkomt niet aan den indruk, dat hier een hoogst merkwaardige wisselwerking tusschen wetenschap en industrie heeft plaats gevonden, een wisselwerking, die onze Technische Hoogeschool tot groote dankbaarheid moge stemmen.

Wie zich in de wetenschappelijke jeugdjaren van den jubilaris van heden tracht te verplaatsen, ontmoet daar in eerste instantie den man, die in 1872 op één en twintigjarigen leeftijd het diploma van technoloog aan de Polytechnische School verwierf. Welke omstandigheden deze studierichting hadden bepaald, laat zich niet met zekerheid zeggen, niet onwaarschijnlijk heeft de bijkomstige omstandigheid, dat een oom van den jubilaris eenige jaren als hoogleeraar voor de waterbouwkunde aan de Koninklijke Akademie werkzaam was geweest, er toe bijgedragen de keuze der familie op Delft te doen vallen. En wellicht waren het vóór alles pecuniaire overwegingen, welke beslisten, dat de jonge MARTINUS WILLEM niet overeenkomstig een eerbiedwaardige familietraditie zich voor het diploma van civiel ingenieur zou bekwamen, doch de kortere en mitsdien minder kostbare opleiding voor technoloog zou volgen.

Hoe dit ook zijn moge, één ding wel staat vast, het was geen roeping, die BEIJERINCK technoloog deed worden. Duidelijk toch blijkt dit uit het feit, dat zijn eerste daad na het behalen van het diploma was een gang naar Minister FRANSEN VAN DE PUTTE, met het verzoek om als houtvester bij het Boschwezen in Nederlandsch-Indië te worden geplaatst. De liefde voor de levende natuur, waarvan ook reeds zijn jongensjaren getuigen, was hem blijkbaar te machtig geworden en deed hem alles beproeven om aan de hem toegedachte loopbaan van industrieel scheikundige te ontkomen.

BEIJERINCK's vrijmoedige stap schijnt in de oogen van FRANSEN VAN DE PUTTE genade te hebben gevonden, maar spoedige teleurstelling volgt. Bij de onvermijdelijke keuring, wordt BEIJERINCK wegens hartzwakte ongeschikt voor den Indischen dienst bevonden.

Men huivert thans bij de gedachte, welke de gevolgen zouden zijn geweest, indien de medicus de kracht van dit hart, dat tot op den huidigen dag zijn plichten zoo trouw en onafgebroken vervult niet zoo hopeloos had onderschat! Want ook voor BEIJERINCK geldt de overweging, welke ik vóór eenigen tijd in het Amerikaansche tijdschrift „Science" zoo markant vond weergegeven met de volgende woorden:

„DARWIN and LINCOLN were born on the same day. If the two infants had been exchanged there would have been no DARWIN and no LINCOLN. What a man can do is determined by his native equipment, what he does is determined by the circumstances of his life."

De ondervonden teleurstelling vermocht intusschen BEIJERINCK allerminst te ontmoedigen, integendeel was zij voor hem blijkbaar een spoorslag om in de richting zijner keuze voort te gaan. Nog geen jaar later toch vinden wij hem terug als iemand, die aan de Leidsche Universiteit magna cum laude het candidaatsexamen in de planten dierkunde aflegt en die zich daardoor de mogelijkheid verschaft in zijn bestaan te voorzien op een wijze, die althans de gelegenheid openlaat zijne biologische studiën te vervolgen. Eerst als leeraar aan de Landbouwschool te Warffum, later als leeraar aan de Hoogere Burgerschool te Utrecht, dan als leeraar aan de Hoogere Landbouwschool te Wageningen ontwikkelt de botanicus in BEIJERINCK zich op krachtige wijze.

Dan breekt de zoo gedenkwaardige 14de Juni 1877 aan, de dag, waarop BEIJERINCK zijn van nauwgezette waarnemingen en van veel omvattende kennis getuigend proefschrift, alsmede zijn kernachtig geformuleerde stellingen in het openbaar — o.m. ook tegenover de bedenkingen van zijn vriend uit den Delftschen tijd, JACOBUS HENRICUS VAN 'T HOFF — verdedigt. De sierlijke Latijnsche lofrede van zijn promotor SURINGAR laat BEIJERINCK onbewogen — want onbegrepen — over zich heen gaan!

Inmiddels schijnt de kloof tusschen Delft en BEIJERINCK welhaast onoverbrugbaar te zijn geworden. En nog steeds neemt de verwijdering toe in de 7 jaren, die op zijn promotie volgen. Dank zij een reeks van fundamenteel belangrijke verhandelingen, verkrijgt BEIJERINCK snel ook buiten de grenzen van ons land den naam van één van de meest vooraanstaande kenners der plantengallen, zoodat het niet kan verwonderen, dat de Koninklijke Akademie van Wetenschappen hem reeds op 33-jarigen leeftijd onder haar leden opnam.

Maar dan geschiedt het onverwachte. In 1885 ziet Delft BEIJERINCK weer in haar veste terug.

Welke krachten hadden dit wonder bewerkt? Hier past het met dankbaarheid een ander groot man te herdenken. J. C. VAN MARKEN, de geniale leider der in de jaren van tachtig te Delft in opkomst verkeerende industrie der rationeele gist-fabricatie was ook hierin zijn tijd zoover vooruit, dat hij tenvolle de heilzame uitwerking realiseerde, welke wetenschappelijke voorlichting op den gang van zaken in een nijverheidsonderneming kan en moet uitoefenen. Waar het in zijn bedrijf gold een microbe op groote schaal voort te kweeken, besefte VAN MARKEN, dat biologische voorlichting een onmisbaar element voor het welslagen vormde. De gelukkige bemiddeling van een HUGO DE VRIES — met wien BEIJERINCK ook toen reeds door hechte vriendschapsbanden was verbonden — bracht VAN MARKEN tot BEIJERINCK, wiens groote kwaliteiten hij met ware intuitie doorvoelde.

Welke verwachtingen VAN MARKEN van deze uitbreiding van zijn staf koesterde, wordt treffend geïllustreerd door hetgeen het merkwaardige Jaarverslag der Nederlandsche Gist- en Spiritusfabriek over 1884 daarover meedeelt en waarop Dr. F. G. WALLER wel zoo welwillend was mijn aandacht te vestigen. Men vindt daarin namelijk vrijwel in extenso overgenomen een opstel, dat VAN MARKEN kort tevoren aan de komst van den nieuwen „hoofdambtenaar" (!) in de „Fabrieksbode" had gewijd en ik kan de verleiding niet weerstaan hieruit een gedeelte voor te lezen.

Na te hebben uiteengezet, hoe de gist in het bedrijf veelal een strijd op leven en dood voert met bacteriën, gaat VAN MARKEN voort:

„Strijd dus aan de schadelijke bacteriën! Maar een strijd, waarin vuur en zwaard niets vermogen; een strijd, waarvoor de wapens gesmeed worden door den geleerde in het studeervertrek en in het laboratorium. BISMARCK heeft door de overwinningen in den Fransch-Duitschen oorlog het machtige keizerrijk gevestigd, maar lang vóór dien oorlog had de bedachtzame veldmaarschalk VON MOLTKE de zwakheden van den vijand bespied, en in het studeervertrek den veldtocht voorbereid.

Den vijand, die onze gist bedreigt, te leeren kennen; de voorwaarden van zijn bestaan en ontwikkeling te ontdekken; hem in al zijn schuilhoeken te bespieden: in het geil en in de zetgist, in het beslag en in de spoeling, in de lucht binnen en buiten de fabriek; overal hem te vangen en zijn geheimen af te dwingen: ziedaar de voorbereiding waar het in de eerste plaats op aankomt, om hem dan allengs schermutselend terug te dringen, en eindelijk misschien den beslissenden veldtocht te ondernemen. Misschien: want de vijand is zóó klein en zóó onzichtbaar, zóó talrijk en zóó listig, zijn eigenschappen zijn nog zoo weinig bekend, dat de overwinning minst genomen twijfelachtig moet worden geacht.

Een jonge geleerde, maar die zijn sporen op het gebied der natuurwetenschap reeds heeft verdiend, de heer Dr. M. W. BEIJERINCK, heeft het niet beneden zijn wetenschappelijke waardigheid geacht, de taak van een VON MOLTKE in ons nijverheidsbedrijf te aanvaarden. Hij heeft gemeend hier een bij uitnemendheid rijk veld van onderzoek te vinden. Hij verwacht van de navorsching der geheimen, die hier verborgen liggen, hoogere bevrediging — de bevrediging van den ernstigen natuuronderzoeker — dan enkel die van het stoffelijke voordeel, dat wij als een gevolg van zijnen arbeid voor onze onderneming mogelijk achten en waarop wij hopen.

Aan gene zijde van den weg over de villa, naast het in aanbouw zijnde graanpakhuis, wordt een laboratorium gebouwd, voorzien van de meest volkomen mikroskopen en van andere wetenschappelijke werktuigen en inrichtingen. Daar, afgescheiden van het gewoel en gedruisch in onzen rumoerigen bijenkorf, worden den geleerde de rust en de hulpmiddelen aangeboden, die hij voor de vervulling van zijn taak noodig heeft.

Zullen de onderzoekingen practische vruchten voor onze onderneming afwerpen? De heer B. is bescheiden, wetenschappelijk genoeg, om dit vraagteeken voorloopig onbeantwoord te laten staan. Uitdrukkelijk heeft hij dit verklaard, toen hij op mijnen wensch, zich bereid verklaarde de taak te aanvaarden. Wat weten wij nog, na zoovele eeuwen van onderzoek en ontwikkeling, wat weten wij nog van het raadsel, dat leven wordt genoemd? De meest uitstekende geneeskundige staat menigmaal schouderophalend aan het ziekbed van den mensch, die wat hij gevoelt en waar hij lijdt, kan mededeelen en aanwijzen. En hier hebben wij te doen met het leven van wezens, die, met behulp van de meest volkomen instrumenten, nog nauwelijks zijn waar te nemen.

Hoe het ook zij, de komst van een geleerde als Dr. BEIJERINCK is in meer dan één opzicht een belangrijk feit, dat in onzen kring hooge waardeering verdient. Ik wensch volstrekt geen overdreven verwachtingen van zijn werkzaamheid in en voor onze fabriek op te wekken. Maar wel ben ik overtuigd, dat ernstige wetenschappelijke arbeid op het gebied der bacteriologie te eeniger tijd — over één jaar, vijf, tien jaren misschien; wij hebben geloof in de wetenschap en haasten haar niet — te eeniger tijd een enkel straaltje van licht zal werpen in de duisternis van het gistingsbedrijf, en wellicht onberekenbare voordeelen, aan onze onderneming zal kunnen brengen."

Tot zoover VAN MARKEN.

Hier spreekt een ruimte van geest, welke voor dien tijd ongehoord mag worden genoemd, maar welke zelfs heden ten dage nog slechts bij uitzondering bij onze Nederlandsche industrieelen wordt aangetroffen. En deze breede opvatting heeft ook de verdere houding van de Directie der Nederlandsche Gist- en Spiritusfabriek tegenover BEIJERINCK gekenmerkt. Moeilijk anders was dit ook te verwachten van een onderneming, waarbij het experiment zoo in hooge eere was, dat men daar vrijwel van den aanvang af over een proeffabriek beschikte en dit in een tijd, waarop dit begrip elders in den lande nog nauwelijks was doorgedrongen.

Zoo vond de experimentator BEIJERINCK in de Delftsche fabriek een gunstigen bodem om op voort te bouwen en wat ook zijn directe invloed op den gang van zaken in het bedrijf moge zijn geweest, vast staat wel, dat van hem een bevruchtende invloed op zijn omgeving uitging, waarvan de gevolgen buiten twijfel indirect ook voor de uitkomsten der onderneming van groote beteekenis zijn geweest.

Verre van BEIJERINCK aan de engere problemen van het bedrijf te kluisteren, liet de

Directie den grooten onderzoeker alle vrijheid zijn problemen daar te grijpen, waar zijn universeele geest ze vond. Zoo zien wij het verrassend schouwspel, dat uit het fabriekslaboratorium te Delft een stroom van verhandelingen verschijnt, welke in de geheele biologische wereld het grootst mogelijke opzien verwekken. Herinnerd zij hier slechts aan de isoleering van den verwekker der wortelknolletjes der Leguminosen, de uitvoerige studiën over de stofwisseling der lichtbacteriën, de eerste geslaagde pogingen om groenwieren en de gonidiën der korstmossen rein te cultiveeren en deze zoodoende voor het stofwisselingonderzoek toegankelijk te maken, de ontdekking van de ongemeen belangwekkende gistsoort *Schizosaccharomyces octosporus*, enz., enz.

Slechts bij eerstgenoemde ontdekking, eene van de eerste grootte, moge nog even worden stilgestaan.

In BEIJERINCK's proefschrift treft heden ten dage een simpel zinnetje: „Slechts in weinige gevallen zijn de gallen nauwkeuriger, de daartoe behoorende parasieten minder goed bekend; dit is het geval met de wortelknolletjes der Papilionaceën." Wanneer men nu weet, dat 10 jaren nadat dit geschreven werd, de Engelsche onderzoeker WARD het in hooge mate waarschijnlijk maakte, dat de onbekende parasiet een bacterie was, dan kan het niet verwonderen, dat BEIJERINCK, die de unieke combinatie van cecidioloog en bacterioloog in zich vertegenwoordigde, niet rustte alvorens hij ook dezen „galverwekker" in handen had. BEIJERINCK slaagde hier, waar talrijken vóór hem faalden.

Hoe belangrijk deze daad, uit drang naar zuivere wetenschap geboren, voor de praktijk, van den landbouw is geweest, wordt treffend gedemonstreerd door het feit, dat dit jaar nog één enkel Amerikaansch laboratorium — en er zijn er daar vele, die dit werk verrichten — in enkele weken tijds 100.000 cultures van deze door BEIJERINCK voor het eerst geïsoleerde bacterie aan den landbouw afleverde. Eén van de gronden, waarop aan BEIJERINCK de EMIL CHRISTIAN HANSEN-medaille werd verleend, luidde dan ook: „en reconnaissance de sa culture du *Bacillus radicicola*, qui a eu une importance éminente pour le développement et la propagation de la culture des Légumineuses."

Hoe weinigen weten intusschen, dat deze voor den landbouw zoo gewichtige vondst afkomstig is uit een fabriekslaboratorium in een oord, dat door BEIJERINCK zelf in botanisch opzicht met een woestijn is gelijkgesteld!

Inmiddels nam het groeiproces van den onderzoeker BEIJERINCK geleidelijk dergelijke afmetingen aan, dat men besefte, dat het niet verantwoord was hem langer binnen de omgrenzing van het Delftsche fabriekscomplex te houden. Naar alle waarschijnlijkheid is het niet in de laatste plaats wederom aan VAN MARKEN's invloed bij de Regeering te danken geweest, dat deze in 1895 er toe overging BEIJERINCK als hoogleeraar in de bacteriologie aan de Polytechnische School te Delft te verbinden en hem een nieuw te bouwen laboratorium ter beschikking te stellen. De invoering van dit onderdeel der biologie als leervak aan een inrichting van technisch hooger onderwijs was toch op dien tijd zonder antecedent. Sedert zijn tal van andere landen hierin gevolgd, maar zelfs in een dit jaar verschenen Duitsch studiewerk wordt er nog over geklaagd, dat in Duitschland de ontstane achterstand nog nimmer ten volle is ingehaald.

Zoo zijn wij dan de phase van BEIJERINCK's werkzaamheden eerst aan de Polytechnische School, later aan de Technische Hoogeschool, genaderd. Lang daarbij stil te staan, zou om verschillende redenen geen zin hebben. In de eerste plaats bevinden zich onder mijn gehoor toch velen, die hem daarbij gedurende een lange reeks van jaren hebben kunnen gadeslaan en die dus meer dan ik bevoegd zouden zijn BEIJERINCK's beteekenis voor onze Hoogeschool te schetsen. Maar voorts heeft de overgroote meerderheid Uwer het voorrecht genoten — een voorrecht dat ik zelf heb moeten missen — om thans zes jaren geleden uit den mond van den Voorzitter van het huidige Comité een meesterlijk overzicht te verkrijgen van de belangrijkste wetenschappelijke vondsten en ontdekkingen, waarvan het Laboratorium aan de Nieuwe Laan in de 26 jaren van BEIJERINCK's hoogleeraarschap getuige was. En voor diegenen, waarbij de verkregen indrukken mochten zijn verflauwd, kan naar den 2den jaargang van het Vakblad voor Biologen worden verwezen, waarin men deze rede in haar geheel vindt afgedrukt.

Loonender lijkt het daarom de vraag in beschouwing te nemen, hoe reageerde de immer voortschrijdende wetenschap in de ruim zes jaren, welke sedert de grootsche huldiging in 1921 zijn verloopen op BEIJERINCK's werk.

De reacties, welke een groot wetenschappelijk onderzoeker op zijn werk ondervindt zijn van uiteenloopenden aard. Eenerzijds dragen deze een persoonlijk karakter en zijn het de bescheiden middelen, waarover de mannen der wetenschap beschikken om uiting te geven aan de gevoelens van waardeering en bewondering, die het werk van een hen veelal persoonlijk onbekenden medestrijder inboezemt.

Vragen wij ons af, hoe het BEIJERINCK in dit opzicht is vergaan, dan treft ons het merkwaardige feit, dat de groote stroom van eerbewijzen hem juist bereikt in den tijd na den ingang van het emeritaat, toen hij, door zich in het landelijke Gorssel terug te trekken, zijn contact met de wetenschappelijke wereld tot een minimum trachtte te reduceeren. Zoo wordt hem in 1922 de EMIL CHRISTIAN HANSEN-medaille verleend, waarvan de bijgevoegde opdracht o.m. de handteekeningen van een CALMETTE, een TH. SMITH, een SÖRENSEN draagt en welke een waardig pendant vormt van de hem reeds in 1905 door de Koninklijke Akademie van Wetenschapppen verleende LEEUWENHOEK-medaille. Voorts geniet BEIJERINCK de zeldzame onderscheiding van het ,,Foreign Membership" van de Royal Society te London, terwijl ook de Deensche en de Russische Akademie van Wetenschappen hem tot buitenlandsch lid benoemden. Zoo ook de British Society for Medical Research. Verder is hij correspondeerend lid van de ,,Society of American Bacteriologists", van het Tsjecho-Slowakisch Botanisch Genootschap te Praag en van de ,,Deutsche Boden-Gesellschaft", terwijl de ,,Société microbiologique à Leningrad", de ,,Wiener Gesellschaft für Mikrobiologie" en de ,,Société pour la zymologie pure et appliquée à Bruxelles" hem alle tot haar eerelid benoemden. Eindelijk was hij Honorary Chairman van het verleden jaar te Ithaca gehouden ,,International Congress of Plant Sciences", terwijl hem tevens aan het Serumlaboratorium der Vee-artsenijkundige en Landbouwkundige Hoogeschool te Kopenhagen een honoraire positie werd verleend. Zoo zien wij nog na 1921 Engelschen, Duitschers, Belgen, Denen, Tsjechen, Oostenrijkers, Russen en Amerikanen, microbiologen, botanici, medici, veterinairen en bodemkundigen op BEIJERINCK's werk reageeren. Met opzet vermeldde ik hier al deze onderscheidingen, omdat de overgroote meerderheid nimmer tot de dagbladpers doordrong.

Maar uit het feit, dat deze stroom van onderscheidingen hem eerst in Gorssel bereikte zijn twee dingen af te leiden. Eenerzijds blijkt er uit, hoe BEIJERINCK in zijn geheele rustelooze onderzoekersbestaan de propaganda voor zijn wetenschap, voor zijn eigen machtigen geest, steeds heeft verwaarloosd, anderzijds hoe de grootsche daad van het huldigingscomité van 1921, dat de verspreide geschriften van BEIJERINCK in 5 forsche deelen verzameld liet herdrukken, beantwoord heeft aan de bedoeling, namelijk de wetenschap te wijzen op schatten, waaraan zij voorbijging.

Maar de wetenschappelijke onderzoeker kent naast de bovengeschetste reacties, ook reacties van anderen aard. Deze bestaan hierin, dat hij mag vaststellen, dat door hem verkregen uitkomsten niet altijd zijn kanteelen op het trotsche gebouw der wetenschap, maar fundamenten die het verrijzen van nieuwe grootsche vleugels mogelijk maken. Ook deze reacties, van hooger orde nog dan de eerder genoemde, zijn BEIJERINCK ruimschoots ten deel gevallen. Dit uitvoerig te documenteeren, zou mij te ver voeren; slechts enkele voorbeelden wil ik U daarom noemen.

Hoort dan hoe in de aan de HANSEN-medaille toegevoegde opdracht, naar aanleiding van BEIJERINCK's ontdekking van het merkwaardige micro-organisme, dat als het belangrijkste agens van de stikstofverrijking van den braakliggenden akkerbodem mag worden beschouwd, wordt getuigd: ,,En reconnaissance de sa découverte de l'*Azotobacter chroococcum* dont les propriétés biologiques particulières ont été largement mises à profit dans les recherches pratiques sur le sol".

Maar de bewuste opdracht noemt behalve de twee reeds genoemde nog een derde speciale motiveering voor het verleenen der onderscheiding. Zoo heet het daar ook nog: ,,en reconnaissance de la fondation du principe de l'application des méthodes électives pour isolement des microbes".

Hoe bevruchtend het principe der electieve cultuur, der ophoopingsmethode, zooals BEIJERINCK haar noemt, op de geheele ontwikkeling der microbiologie heeft gewerkt, laat zich niet onder woorden brengen.

Door BEIJERINCK werd voorts het eerst een helder licht geworpen op de bacteriesoort, welke de zeldzame eigenschap bezit om sulfaten tot zwavelwaterstof te reduceeren en welke daardoor als één der hoofdschuldigen moet worden beschouwd van den stank der verontreinigde stadsgrachten. Maar ook haar beteekenis als factor in het geologisch gebeuren werd reeds door BEIJERINCK aangeduid en het moet ongetwijfeld een groote voldoening voor BEIJERINCK zijn, dat thans — 32 jaren na het verschijnen zijner verhandeling — verschillende publicaties bewijzen, dat ook geologen in toenemende mate van de beteekenis van *Vibrio desulfuricans* voor verschillende hunner problemen doordrongen geraken.

Ten slotte moge in het beschouwde verband nog van één verhandeling melding worden gemaakt, namelijk die over de mozaïkziekte van de tabak, waarin BEIJERINCK één der grondproblemen der biologie, te weten de vraag naar de eenvoudigste gedaante, waarin het leven zich manifesteert, aanroert. Er is heden ten dage zeker geen schooner getuigenis mogelijk voor de beteekenis van BEIJERINCK's stoutmoedige conceptie van het ,,contagium vivum fluidum" dan de hieronder volgende woorden van den beroemden ontdekker van den bacteriophaag, FÉLIX D'HÉRELLE, woorden welke deze in 1925 — vijf en twintig jaren na het verschijnen van BEIJERINCK's verhandeling — sprak, toen hij de LEEUWENHOEK-medaille der Koninklijke Akademie van Wetenschappen in ontvangst nam. D'HÉRELLE zeide dan bij deze gelegenheid o.m.: ,,On a beaucoup discuté la conception de BEIJERINCK, mais je ne pense pas qu'on en ait saisi toute la profondeur. Toute la biologie reposait, repose encore, sur l'hypothèse fondamentale que l'unité de matière vivante, c'est la cellule. BEIJERINCK le premier, s'est affranchi de ce dogme et a proclamé de fait, que la vie n'est pas le résultat d'une organisation cellulaire, mais dérivé d'un autre phénomène, qui ne peut dès lors résider que dans la constitution physico-chimique d'une micelle protéique."

Het wil ook mij voorkomen, dat de gedachten, welke BEIJERINCK in zijn beroemde voordracht over: ,,De infusies en de ontdekking der bacteriën" aan zijn ,,contagium vivum fluidum" wijdt, voorbestemd zijn nog een belangrijke rol in den komenden eindstrijd over den aard van den bacteriophaag te spelen, zij het dan ook, dat het mij niet uitgesloten lijkt, dat het door D'HÉRELLE gehanteerde wapen zich wel eens tegen hem zelf kon keeren.

Genoeg intusschen, het voorafgaande zal voldoende zijn geweest om U den indruk te geven, dat aan BEIJERINCK in de laatste jaren ook voldoeningen van hooger orde ruimschoots zijn ten deel gevallen.

Hooggeachte aanwezigen,

Door op deze samenkomst het woord te voeren, beoogde ik intusschen niet alleen de beteekenis van BEIJERINCK voor de wetenschap en daarmede ook voor onze Hoogeschool op dit oogenblik U nog eens levendig voor oogen te brengen. Ik had daarmede nog een tweede doel en dit is U, Mijnheer de Voorzitter van het College van Curatoren, nog eens den warmen dank te betuigen, niet alleen van mijzelf, maar ook van allen, wien het welzijn van het Laboratorium voor Microbiologie ter harte gaat, dat Gij er wel in hebt willen toestemmen, dat het heden aan de Technische Hoogeschool aangeboden geschenk zijn blijvende bestemming vindt in het onder mijn beheer staande gebouw.

Hoezeer BEIJERINCK met dit gebouw was samengegroeid, blijkt wellicht nog duidelijker dan uit zijn destijds gedane weigering om het te verruilen voor een mogelijk wijdere perspectieven biedend instituut te Leiden, uit het feit, dat hij zoo hardnekkig weigert er terug te komen, sinds de harde wet hem er van scheidde.

Zeker de hoogleeraar BEIJERINCK was van de Technische Hoogeschool in haar geheel, maar het laboratorium aan de Nieuwe Laan was een stuk van BEIJERINCK zelf en het is dan ook passend, dat daar in de eerste plaats de herinnering aan den grooten geleerde blijft voort leven.

Gaarne geef ik U hier de verzekering, dat wij de plaquette van heden af aan zullen hoeden en bewaken als ons kostbaarste bezit.

Wanneer men leest, dat in Amerika weer nieuwe millioenen zijn bijeengebracht voor een instelling van wetenschap of onderwijs, dan hoort men somtijds de vraag opwerpen, of men deze nu zal besteden voor ,,bricks" dan wel voor ,,brains". Met deze vraag wil men dan uiting geven aan het besef, dat de uitkomsten der te scheppen instelling niet alleen afhankelijk zijn van een kostbaar gebouw en dito materieele uitrusting, doch dat daarnaast ook ,,hersenen" d.w.z. jonge intelligente werkers worden vereischt.

Maar het komt mij voor, dat ook een dergelijke uitspraak nog een miskenning inhoudt van de voorwaarden, welke een noodzakelijkheid zijn voor het welslagen eener dergelijke instelling. En dat onmisbare is iets, wat voor geen geld te koop is in deze wereld, namelijk een direct tot het gemoed der werkers sprekend voorbeeld van ongebreidelde toewijding tot, ja volledige overgave aan, onderzoek en wetenschap.

Dit voorbeeld vinden wij microbiologen in BEIJERINCK. Immers alleen een dergelijke overgave kon hem op vijf en zeventigjarigen leeftijd nog de blijkbaar uit het diepst van zijn gemoed opwellende woorden doen schrijven, woorden, welke sedert aan den wand van zijn oude laboratorium prijken:

,,Gelukkig zij, die nu beginnen".

Van deze overgave zien wij in zijn bronzen beeltenis het symbool. Moge dit symbool en het daardoor opgewekte besef van op gewijden bodem te werken er toe bijdragen, dat iets van BEIJERINCK's liefde voor de wetenschap, voor *zijne* microbiologie, ook op komende generaties worde overgedragen.

Appendix J.

Interview with BEIJERINCK published by Mrs. W. VAN ITALLIE–VAN EMBDEN. *)

In Gorssel aan 't station. De hotel-auto wacht. Prof. noodt mij binnen. We tuffen 't dorp uit. Aan den prachtigen landweg 'n eenvoudig buitenhuis.

'k Word 'n kamer binnengeloodst om wat te rusten na de reis. In den voortuin hoor ik Prof. redeneeren met 'n reiziger in stofzuigers. Zelfs 't klemmend argument: „We hebben er al een" bleek ter afwering niet voldoende. Na lang praten: reiziger àf. Prof. knikt, fier op zijn overwinning, naar boven.

Wat 'n apart gezicht! Iets van 'n ouden leeuw in bouw en dwang van wil. Zal wel 'n éénling zijn in deze goedige boerenmenschen-streek!

De studeerkamer is uiterst sober. Prof. zet 'n mutsje op, en 'n bril. Blijft tòch een leeuw, die voor grootvader speelt!

„Is de temperatuur hier naar uw zin, mevrouw? precies 19°."

„Dus hier woont: „de Hollandsche bacteriejager". Wat 'n lang leven van werk-in-wetenschap overziet u. Erfde u van uw ouders dien onderzoekersdrang?"

„Beste menschen. Niet mijn studie-aanleg. Mijn grootvader wel. 'k Zit altijd te piekeren in dingen van de erfelijkheidsleer. Heel jong al werkte 'k me in, in DARWIN. Vond toèn al steun in hèm bij mijn botanisch-zoölogische vorschingen. Daarheen dreef mijn diepste natuur. — 't Leven kneedde me tot chemicus."

„Was uw vader 'n gestudeerde?"

„Nee, handelsman. En daarvoor niet geschikt. 'n Sociale tragedie; Vader had 'n kunstenaarsaanleg; niet dwingend genoeg om dóór te zetten. Ach, moeilijke jaren thuis. . . 'k Ging naar Haarlem op de eerste H.B.S. Konden mijn ouders, met veel opoffering, nog net bekostigen."

„Voèlden ze uw aanleg?"

„Weet ik niet. In de 3de klas kreeg ik den Eersten Prijs van de Holl. Maatschappij voor Landbouw. Had 'n herbarium ingezonden van 150 planten: de flora van Kennemerland. Waren onbekende soorten bij."

„Wijst wèl op aanleg!"

„Dat besliste. 'k Zei: 'k ga studeeren. Moeilijkheden? Gebeuren zàl het. 'n Oom deed me op de Polytechnische School in Delft. Goed bedoeld, en fout. 'k Hoorde in Leiden, voor de botanie. Kwam in Delft te staan naast VAN 'T HOFF. Die hielp me. De hoogleeraar gaf goed college, bemoeide zich verder niet met de studenten. 'k Deed al gauw, wat ik wou. Kwam slecht op 't lab. Maakte preparaten voor mijn pleizier. VAN 'T HOFF deed mee, máár. . . hij verwaarloosde ook 't andere niet. Had sterker plichtsgevoel. — Zijn verloren jaren voor me geweest. Had eind-examen gymnasium moeten doen."

„Zou u dan geen moeite hebben gehad met de klassieke talen?"

„Moeite zéker. Zou er gekòmen zijn: de weg naar de botanie."

„Was u zich uw aanleg niet bewust?"

„Te kinderlijk. Wat mijn ouders vonden, was 'n natuurwet. Deed eindexamen als technoloog. VAN 'T HOFF was al weg. 'n Buitengewoon voortreffelijk student. — Op mijn vorming was zijn invloed niet altijd gunstig; 'k dacht: wat ben ik. . . vergeleken met hèm. Hij had toèn al ontdekkingen van waarde gedaan. 'k Nam intensief deel. We praatten, we werkten dagen en nachten."

„Hoe kunt u dan spreken van verloren jaren?"

Bedachtzaam wegend: „'k Ben in twijfel." — Vast de conclusie: „De andere weg was béter geweest. — 'k Moest gauw gaan verdienen. 'k Vroeg mijn ouders: „Geef me één jaar in Leiden." THORBECKE had gezorgd; als technoloog kon je daar examen doen in de plant- en dierkunde."

„Dus tòch bereikt!"

„Niet grondig. Te kort."

---

*) Reprinted from "De Groene Amsterdammer" of March 17th, 1928.

'n Melancholie, als 'n sluier, legde zich over 't gezicht.

„Werd leeraar in Warffum. Hoofdvak: plantkunde. Ja, tòen kon ik tòch nog aan mijn liefde toegeven!"

De sluier vaagde weg. De wil sprong uit de oogen.

„Kon leeraar worden in Amsterdam. Weigerde. Was niet rijp. 'n Jaar later: Utrecht. Heb 't gewáágd. Ontmoette daar wéér VAN 'T HOFF."

„Voelde hij voor u?"

„Wel iets . . . van uit zijn hoogte." — Trotsch: „'k Was niet in àlles zijn mindere; 't zij met bescheidenheid gezegd, of ònbescheiden, gelijk u wilt. 'k Was beter bioloog. Bouwde 'n algemeene theorie op. Hij zei: „Die is onhoudbaar." 'k Voelde: dat oordeel is verkeerd. . . Nu, nu aan 't eind van mijn leven, zie ik mijn fout: 'k bezat niet genoeg eerzucht. VAN 'T HOFF zei fier: „Ambition is my idol." Als ik dien trek had gehad, zou ik ook wel eenigen roem. . ."

„De geleerden zeggen: dien hèbt u. De hóógste onderscheidingen zijn u „toegestroomd."

„Eerst in de laatste jaren. En dat pleit voor mijn stelling. Hòe publiceerde ik? Kon me niet schelen waarin: 't eerste tijdschrift dat me in handen kwam. Werd soms heel weinig gelezen. 'k Dacht: 't is tòch niet de moeite waard. — Ziedaar mijn fout: ik had moeten denken: 't Is het wèl. Anderen mochten anders oordeelen. Ik niet."

„Is u nooit getrouwd geweest?"

„Ging te veel op in mijn werk."

'n Schampere glimlach van zelfspot verbreedt den mond, die gesloten blijft.

„U promoveerde op: „Bijdrage tot de morphologie der plantegallen."

„'k Was tamelijk handig in 't gebruik van de microscoop. Had 't mezelf geleerd. — In Utrecht, als leeraar, in zwaren tijd. Groote klassen."

„Men prijst u: 'n geboren docent."

„Was ik niet voor de H.B.S."

„U is te bescheiden."

Verbaasd: „Bescheiden? 'k Was 'n geboren Professor. Kwam al uit in Wageningen. Was 9 jaar werkzaam aan de Hoogere Landbouwschool. Met pleizier, en vruchtbaar. Maar 'k bezat geen eerste klas laboratorium. 'k Ging naar den Minister. 'k Had grootsche plannen: wou de cultuur van onze granen verbeteren; zou ook op de praktijk invloed hebben gehad. — 'k Had in mijn eentje MENDEL weer ontdekt: 5 jaar vóór HUGO DE VRIES. — De Minister begreep niets van mijn betoog. Bleef ijskoud. Hielp niet. . . Toen ging ik over naar de Gistfabriek van VAN MARKEN in Delft. Had me in Wageningen moeten vàstbijten. Quand même."

„'t Is 'n eigenschap van den mensch om zwaar te wegen wat hij niet bereikt en te licht te tellen wat hij heeft volbracht."

„Volkomen juist. 'k Mag dat niet wegcijferen: de gistfabriek heeft me veel werk gegeven, en, 'k zal 't niet ontkennen: vruchtbaar werk. In 'n eigen richting: de microbiologie. — Hier hebt u 't portret van VAN MARKEN. Wat 'n nobel gezicht! 'n Héél ander karakter dan ik. Mij konden sociale toestanden niet schelen. Een zware fout. Als ik meer aangeboren gevoel had gehad voor mijn medemenschen zou mijn leven innerlijk rijker zijn geweest. — 'k Zag alléén: de wetenschap. Kan er niets aan doen."

„En op één s mocht u zich geheel aan haar geven: als Professor in de bacteriologie."

„'n Prachtig laboratorium! 't Eerste in de wereld waar dàt deel van de biologie tot 'n eigen leervak werd geheven. En zulke knappe, jonge medewerkers. Door hèn ben ik wetenschappelijk frisch gebleven. Toen ik 70 jaar werd, hebben vrienden mijn „Verzamelde Geschriften" uitgegeven."

„Vijf zware, kloeke deelen."

„Dáároan schrijf ik 't toe, dat ik nog zoo bekend geworden ben."

„Uw leerlingen kwamen van heinde en ver."

„Veel snuiters uit Midden-Europa. Ook Engelschen, Amerikanen."

„Dan was u al toch wèl bekend vóór uw zeventigste!"

„'k Had naam door 'n onderzoek over lichtgevende bacteriën. Eén geslaagde proef brengt je de wereld door."

„Hebt u veel gereisd?"

„'k Ben twee keer bij PASTEUR geweest. Van zijn kostbaren tijd gestolen. Verkeerd. En ook voor den bezoeker . . . wat brengt 't dan: streeling van ijdelheid? Bestudeer hun werk thuis. Doe hun proeven na. Ontdek je zèlf. Ik deed 't te laat. Eerst drie jaar vóór mijn aftreden had ik begrepen, hòe ik moest doceeren. Had gevonden: dè cursus voor de micro-biologie. U kunt dat kale pedanterie noemen; ik voel 't als waarheid."

„Om een goed inzicht in uw waarde te krijgen, moet je lezen wat ànderen over u schrijven. U is wel een zeer bijzondere mengeling van trots en nederigheid. — Uw werken zijn in veel talen vertaald."

„Kunt u dat lezen? Kijk eens: 't is Russisch. Alleen aan de begeleidende figuren herken ik mijn stuk. Over 'n bacterie die het giftige kool-oxyde op eet. Die ontdekking heeft me toch zoo mal beroemd gemaakt. — Nu ja, mijn werk heeft wel wàt invloed gehad. Ook op de praktijk. Ach, ten slotte had die 't óók wel gevonden. . . zonder mijn geleerdheid."

„'t Is toch heerlijk als de prakticus wéét wat hij doet."

„Weet? Ook de wetenschap weet niet. . . de laatste oorzaken. 't Allerdiepste blijft onontdekt. Ziet u deze cultuur: wit, blauw, zwart. Van dezelfde microben. Waarom in drie kleuren? Wat is de oorzaak van die variabiliteit? Weet niet."

„Een ander zal weten, staande op ùw schouders."

„'t Oerbegin blijft Mysterie. — Mag ik u mijn tuin eens laten zien? De zon schijnt; de bloemen staan zoo mooi."

Appendix K.

Obituary articles.

W. B(ULLOCH), Martinus Willem Beijerinck. Proceedings of the Royal Society, Ser. B, **109,** I, 1932.

G. VAN ITERSON JR., Martinus Willem Beijerinck. Nieuwe Rotterdamsche Courant van 5 Januari 1931 (Avondblad).

G. VAN ITERSON JR., Martinus Willem Beijerinck. Berichte der deutschen botanischen Gesellschaft, Jahrgang 1934, Band **52,** 2. Generalversammlungs-Heft. S. 115.

A. J. KLUYVER, In Memoriam Prof. M. W. Beijerinck. De Telegraaf van 5 Januari 1931 (Avondblad).

A. J. KLUYVER, In Memoriam Prof. Dr. M. W. Beijerinck. Nederlandsch Tijdschrift voor Hygiëne, Microbiologie en Serologie **5,** 173, 1931.

J. J. VAN LOGHEM, Beijerinck en de kennis der bacterieele veranderlijkheid. Nederlandsch Tijdschrift voor Geneeskunde **75,** 1046, 1931.

A. MAYER, Der holländische Botaniker, Bakteriologe und Biologe M. W. Beijerinck. Die Naturwissenschaften **19,** 302, 1931.

JAN SMIT, Beijerinck's levenswerk. Chemisch Weekblad **28,** 94, 1931.

SELMAN A. WAKSMAN, Martinus Willem Beijerinck 1851–1931. Soil Science **31,** 245, 1931.

SELMAN A. WAKSMAN, Martinus Willem Beijerinck. The Scientific Monthly **33,** 285, 1931.

F. A. F. C. WENT, In Memoriam M. W. Beijerinck. Verslag Koninklijke Akademie van Wetenschappen, Afdeeling Natuurkunde **40,** 6, 1931.

GPSR Compliance

The European Union's (EU) General Product Safety Regulation (GPSR) is a set of rules that requires consumer products to be safe and our obligations to ensure this.

If you have any concerns about our products, you can contact us on ProductSafety@springernature.com

In case Publisher is established outside the EU, the EU authorized representative is:

Springer Nature Customer Service Center GmbH
Europaplatz 3
69115 Heidelberg, Germany

Batch number: 10585060

Printed by Printforce, the Netherlands